International Science

Coursebook 3

International Science

Coursebook 3

Karen Morrison

HODDER
EDUCATION
AN HACHETTE UK COMPANY

The Publishers would like to thank the following for permission to reproduce copyright material:

Text credits
p. 56 "Heal' the world's biodiversity', used with permission of The Associated Press Copyright © 2009. All rights reserved.

Photo credits
All photos supplied by Mike van der Wolk (mike@springhigh.co.za) except for **p.8** Fig 2.2*l* © Louise Murray/Alamy, *r* © F1 Online/Rex Features; **p.31** Fig 4.6 © 2009 John Warburton-Lee Photography/photolibrary.com; **p.64** Fig 7.4 © Ria Novosti/Science Photo Library, Fig 7.5 © Lawrence Berkeley National Laboratory/Science Photo Library; **p.113** Fig 12.1 © Antti Aimo-Koivisto/Rex Features; **p.120** Fig 12.11 © 2009 Oxford Scientific/photolibrary.com; **p.123** Fig 13.1 © Peter Menzel/Science Photo Library.

Every effort has been made to trace all copyright holders, but if any have been inadvertently overlooked the Publishers will be pleased to make the necessary arrangements at the first opportunity.

Hachette UK's policy is to use papers that are natural, renewable and recyclable products and made from wood grown in sustainable forests. The logging and manufacturing processes are expected to conform to the environmental regulations of the country of origin.

Orders: please contact Bookpoint Ltd, 130 Milton Park, Abingdon, Oxon OX14 4SB.
Telephone: (44) 01235 827720. Fax: (44) 01235 400454. Lines are open 9.00–5.00, Monday to Saturday, with a 24-hour message answering service. Visit our website at www.hoddereducation.co.uk.

© Karen Morrison 2009
First published in 2009 by
Carmelite House
50 Victoria Embankment
London EC4Y 0DZ

Impression number	10	9	8
Year	2020	2019	2018

Cover photo © TEK Image/Science Photo Library
Illustrations by Robert Hichens Design and Macmillan Publishing Solutions
Typeset in 12.5/15.5 pt Garamond by Macmillan Publishing Solutions
Printed in Dubai

A catalogue record for this title is available from the British Library

ISBN 978 0 340 96602 0

Contents

Chapter 1 Testing for substances

⬆ **Figure 1.1** This pupil is testing whether the leaf contains starch.

Scientists often need to carry out standard tests to find out what a substance is. For example, they might test food to see if it contains starch, protein or fat. If a gas is given off in a reaction, scientists might test it to see what kind of gas it is.

In this chapter, you will learn how to:

- test for starch, protein and fat
- test for oxygen, carbon dioxide and hydrogen.

Unit 1 Testing for food substances

Last year, you learned that our food contains carbohydrates (starches and sugars), proteins and fats. Do you remember which foods are good sources of each of these nutrients?

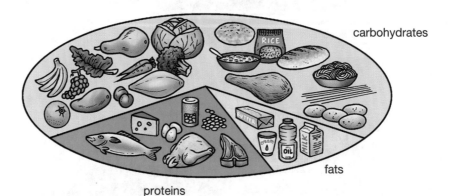

carbohydrates

fats

proteins

→ Figure 1.2
Foods in the three
main food groups

Food tests

In science, you sometimes need to test foods to see if they contain carbohydrate, protein or fat. The table summarises the standard tests you can use. Read the information carefully. When you learn about photosynthesis in Chapter 2, you will use some of these tests to find out if parts of plants contain certain substances.

Food group	Test	What you do	Results
starch	iodine test	Add a few drops of iodine solution to the food.	A blue–black colour means starch is present. Brown means no starch.
simple sugars	Benedict's test	Mash food and mix it with water. Put the mixture into a test tube and add an equal amount of Benedict's solution. Heat the mixture in a water bath.	Orange–red, yellow or green means sugar is present. Blue means no sugar.
lipid (fat or oil)	grease spot test	Rub a small piece of the food onto filter paper. Hold the paper up to the light to check for a translucent mark.	A translucent greasy mark means the food contains lipids.
lipid (fat or oil)	emulsion test	Dissolve the food in ethanol. Pour the solution into a clean test tube of water.	Water turns white and cloudy if lipid is present. It stays clear if there is no lipid.
protein	Biuret test	Mash the food and mix it with water. Add drops of potassium hydroxide until the solution clears. Then add a few drops of dilute copper sulphate. Shake mixture.	A violet or purple colour shows that there is protein in the food.

When you are testing several different foods, you must not mix them up, or your results will not be reliable. So keep the foods in separate dishes, use separate knives to cut them up and wash the test tubes after each test. For most tests, you need to mash up the food, or cut it into small pieces, before you can test it.

Experiment
1.1

Testing food substances

Aim

To test different foods to see what nutrients they contain.

You will need:

- samples of fruit juice, banana, bread, milk, egg white, margarine, breakfast cereal, and sweet biscuits or cake
- pieces of two other foods that you eat regularly
- equipment for testing (your teacher will supply this)

Method

1 Divide the food samples into groups depending on the main substance you think each contains – carbohydrate, protein or lipids.
2 Decide which tests you will perform on each sample.
3 Carry out the tests.
4 Use a table like this to record your results.

Food sample	Test	What we noticed (observations)	What this tells us (deductions)

Activity 1.1   **Answering questions about food tests**

Food tests do not tell you how much of each nutrient is found in the food.

1 Why is the amount of each nutrient important?

2 Which nutrients are needed for a balanced diet?

Unit 2 Testing for different gases

In Figure 1.3, magnesium has been placed in weak hydrochloric acid. A **chemical reaction** is taking place. This reaction produces bubbles of hydrogen gas and a solution of magnesium chloride. But how can we be sure the gas is hydrogen?

Many reactions produce gases. In science, it is important for us to know what gas is produced as a result of a reaction. To find out, we collect the gas and **test** it. There are many different tests for gases, but you will only test for oxygen, carbon dioxide and hydrogen.

↑ Figure 1.3
Magnesium reacting in dilute hydrochloric acid

Collecting gases

You can collect the gas from a reaction in different ways. The method you use depends on how dense the gas is, and whether or not it is **soluble** (dissolves) in water.

Study Figures 1.4 to 1.6 carefully to see how you can collect oxygen, carbon dioxide and hydrogen gas in the laboratory.

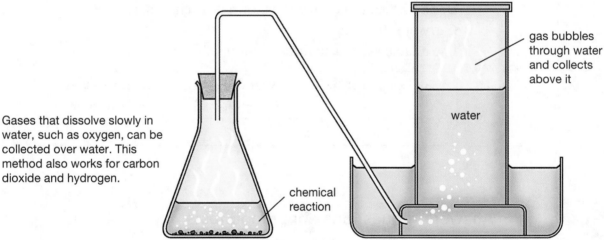

Gases that dissolve slowly in water, such as oxygen, can be collected over water. This method also works for carbon dioxide and hydrogen.

gas bubbles through water and collects above it

water

chemical reaction

↑ Figure 1.4 Oxygen gas can be collected over water.

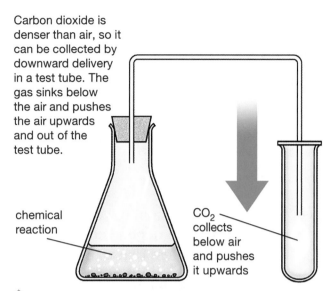

Carbon dioxide is denser than air, so it can be collected by downward delivery in a test tube. The gas sinks below the air and pushes the air upwards and out of the test tube.

chemical reaction

CO_2 collects below air and pushes it upwards

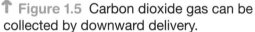

⬆ **Figure 1.5** Carbon dioxide gas can be collected by downward delivery.

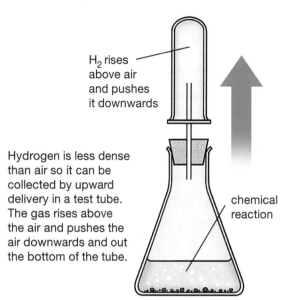

H_2 rises above air and pushes it downwards

Hydrogen is less dense than air so it can be collected by upward delivery in a test tube. The gas rises above the air and pushes the air downwards and out the bottom of the tube.

chemical reaction

⬆ **Figure 1.6** Hydrogen gas can be collected by upward delivery.

Tests for gases

Scientists use the properties of different gases to identify them.

- Oxygen supports combustion – it helps things burn – so we can test for oxygen using a glowing splint of wood. If the splint catches alight, then the gas is probably oxygen.
- Carbon dioxide does not support combustion – it stops things burning. If you place a glowing splint in carbon dioxide, it goes out. Carbon dioxide also reacts with limewater, turning it from a clear liquid to a milky white colour.
- Hydrogen is explosive. If you put a lighted splint into hydrogen, it explodes with a popping sound.

 **Activity 1.2**   **Summarising information in a table**

Copy and complete this table to summarise what you have learned about oxygen, carbon dioxide and hydrogen.

Gas	Is it denser than air?	Is it soluble in water?	Method of collection	Test
oxygen				
carbon dioxide				
hydrogen				

Chapter summary

✓ Scientists use several different standard tests to identify substances.

✓ Benedict's test tells us if a food contains simple sugars.

✓ We use iodine solution to test whether a substance contains starch.

✓ The grease-spot test and the emulsion test can be used to find out if a substance contains lipids.

✓ We use the Biuret test to test for protein.

✓ We can collect the gas from a chemical reaction by bubbling it though water, by upward delivery or by downward delivery. The method we use depends on whether the gas is soluble in water, and whether it is more or less dense than air.

✓ We use a glowing splint to test for oxygen. If the splint catches alight, oxygen is present.

✓ We use a lighted splint to test for carbon dioxide and hydrogen. If the flame goes out, carbon dioxide is present. If the flame makes the gas explode with a pop, then hydrogen is present.

Revision questions

1 Copy and complete this table by ticking the boxes to show what you learned about different foods when you tested them for carbohydrate, protein and lipids.

Food	Contains sugar	Contains starch	Contains protein	Contains lipids
banana				
bread				
milk				
egg white				
margarine				

2 You find a seed from a plant. You wonder if seeds like this could be used to make a cooking oil. How could you test whether the seed is rich in oil?

3 Fill in the missing words in these sentences to summarise what you have learned about testing for oxygen, carbon dioxide and hydrogen gas.
   a) A lighted splint will _____ when it is placed in carbon dioxide.
   b) A lighted splint will _____ when it is placed in hydrogen.
   c) A _____ splint will catch alight when it is placed in oxygen.

Chapter 2 Photosynthesis

↑ **Figure 2.1** In its leaves, this plant is making its own food using sunlight.

You already know that plants are the producers in all food chains. Plants are the only living organisms that can make their own food. Animals have to eat plants or other animals to get the food they need. The process by which plants make food is called photosynthesis.

In this chapter, you will learn more about how plants use sunlight and other substances to produce their own food. You will:

● learn about photosynthesis and write a word equation to describe the process
● carry out tests to show that starch is a product of photosynthesis
● carry out experiments to show that plants need light, chlorophyll and carbon dioxide for photosynthesis to take place
● examine leaves and state their role in photosynthesis
● describe how food substances are transported and stored in plants.

Unit 1 Plants and food

↑ **Figure 2.2** Plants are the producers in all food chains. Animals cannot make their own food – they must eat plants or other animals to get energy.

All living things depend on light energy from the Sun, directly or indirectly. Plants capture the energy in sunlight and lock it away in food (chemical energy), in a process called **photosynthesis**.

Plants can store the food they make. Humans and animals eat plants and then convert the stored food from the plants into energy for their own movement and growth. The process that living things use to release the energy stored in food is called **respiration**.

Plants' leaves take in carbon dioxide (CO_2) from the air and their roots take in water from the soil. Both of these are needed for photosynthesis.

Most plants are green – they have green leaves, they may have green stems and some plants even have green flowers. The green colour comes from **chlorophyll**. Chlorophyll is found in **chloroplasts**. These are tiny structures in plant cells that act like small solar converters.

The chlorophyll absorbs light and uses it to change carbon dioxide (CO_2) and water into glucose. Oxygen (O_2) is produced as a waste product. We can use a word equation to show what happens:

$$\text{carbon dioxide} + \text{water} \xrightarrow[\text{chlorophyll}]{\text{sunlight}} \text{glucose} + \text{oxygen}$$

Testing leaves for starch

The glucose made in photosynthesis is stored in plants in the form of **starch** (a carbohydrate). We can show that leaves contain starch by testing them with iodine solution. When we do this, we have to get rid of the green colour in the leaves first, so that we can see if the iodine changes to blue–black.

Experiment

2.1

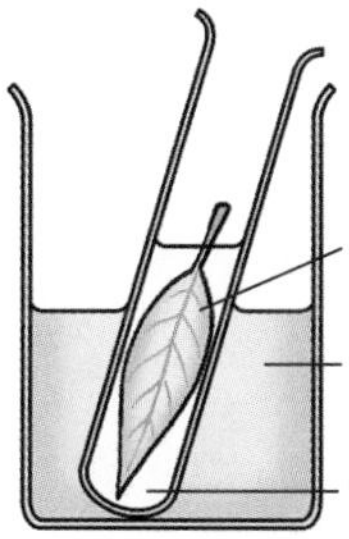

↑ **Figure 2.3** Hot ethanol decolorises the leaf.

leaf
hot water
ethanol

Decolorising a leaf and testing it for starch

Aim

To see whether a leaf contains starch.

You will need:
- a leaf
- a Bunsen burner and tripod
- a beaker or glass jar half-filled with water
- a test tube
- ethanol
- iodine solution
- a white tile
- tweezers

Method

1 Heat the beaker of water over a Bunsen burner, until it boils.
2 Dip the leaf in the hot water for 20 seconds, just long enough to soften it.
3 Turn off the Bunsen burner. Check that everyone else's burners are turned off too.
4 Pour some ethanol into the test tube and put in the leaf. Stand the test tube in the beaker of hot water for 10 minutes, as in Figure 2.3.
5 Rinse the leaf well in water. Your leaf is now ready for the starch test.
6 Place the leaf on a white tile. Place a drop of iodine solution onto the leaf. What happens to the iodine solution?

Activity 2.1   **Explaining photosynthesis**

1 The word photosynthesis comes from the Greek word 'photos', meaning light, and the word 'synthesis', meaning a combination. Explain why you think these words were put together to describe the process that plants use to make glucose.

2 When plants photosynthesise, the chloroplasts absorb light energy.
   a) Name the two compounds that the plants use during photosynthesis.
   b) Name the two products of photosynthesis.

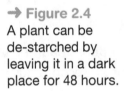 Investigating photosynthesis

In Unit 1, you saw that plant leaves contain starch. But how do we know that plants really produce their own food? In other words, can we prove that the starch is made by photosynthesis and is not just a normal part of leaves?

To show that a plant makes starch during photosynthesis, we must first get rid of any starch in its leaves. If you put a plant in the dark, it cannot photosynthesise. Instead, it uses up all the starch stored in its leaves. This is called 'de-starching' the plant. If you test the leaves after the plant has been in a dark box or cupboard for 48 hours, you will find that they have no starch in them.

➡ **Figure 2.4**
A plant can be de-starched by leaving it in a dark place for 48 hours.

Experiment 2.2

Proving that plants make their own food

Aim

To show that plants produce starch when they are exposed to sunlight.

You will need:
- a potted plant
- equipment for decolorising leaves and testing for starch as in Experiment 2.1.

Method

1 Remove a single leaf from the plant. Decolorise it and test it for starch. What is the result?
2 Put the plant in a dark place for 48 hours.
3 Remove another leaf. Decolorise it and test it for starch. What is the result?
4 Now put the plant in a sunny place for a few hours. Repeat step **3**.

Questions

1 What can you conclude from this experiment?
2 Why do you think it was necessary to put the plant in the dark during this experiment?

Light and photosynthesis

1 Juanita and Samiah started with a de-starched pot plant.

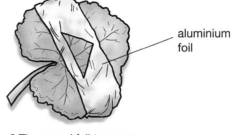

aluminium foil

2 They used foil to cover part of a leaf.

aluminium foil

3 They put the plant in a sunny place for a day.

4 They removed the foil from the leaf and tested that leaf for starch.

➜ **Figure 2.5**
Juanita and Samiah's experiment

Juanita and Samiah carried out the experiment in Figure 2.5. When they tested the leaf for starch, they noticed that:

- the parts of the leaf that were covered by foil stayed brown when the iodine was added
- the parts of the leaf that were not covered by foil turned blue–black when the iodine was added.

So, there was starch in the parts of the leaf that sunlight could reach, but the parts that were in the dark (covered with foil) did not produce starch. Juanita and Samiah concluded that light is needed for leaves to produce starch by photosynthesis.

Activity 2.2   **Answering questions about photosynthesis**

1 What experiment can you use to show that plants produce their own food?

2 How do you de-starch a plant?

3 What happens to the starch when you de-starch a plant?

4 What happens when you put a de-starched plant in sunlight? How do you know this?

5 How can you prove that plants need sunlight for photosynthesis?

Unit 3 What else do plants need?

You have seen that plants need sunlight for photosynthesis. But the word equation on page 8 showed that plants also need water, carbon dioxide and the green pigment chlorophyll for photosynthesis.

Water

Plants cannot make food without water. Water is absorbed through the roots and transported to the leaves through the xylem vessels.

Carbon dioxide

Photosynthesis cannot take place if there is no carbon dioxide in the air. Figure 2.6 shows an experiment that David and Mayura carried out to prove this.

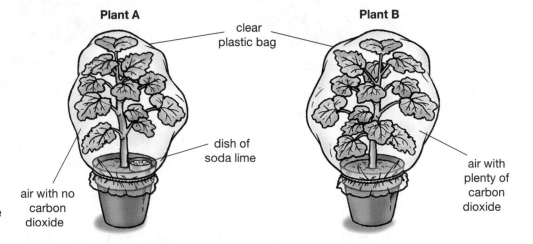

➜ Figure 2.6
An experiment to show that plants need carbon dioxide for photosynthesis

Both plants were de-starched at the start. During the experiment, both plants had enough light and they had the same amount of air. However, David and Mayura placed a small dish of soda lime inside the plastic bag with Plant A. Soda lime absorbs carbon dioxide from the air – so the air around Plant A had no carbon dioxide in it.

After a few hours, the pupils removed the plants from the plastic bags and tested a leaf from each plant for starch.

- The leaf from Plant A remained brown when tested with iodine. This showed that no photosynthesis had taken place.
- The leaf from Plant B turned blue–black when tested with iodine. This showed that starch was present, so photosynthesis had taken place.

The results of the experiment show that carbon dioxide is needed for photosynthesis to take place.

Chlorophyll

Some plants' leaves are not green all over. Figure 2.7 shows some leaves like this. We say that the leaves are variegated. The green parts contain chlorophyll, and the light parts do not.

← **Figure 2.7** Variegated leaves

Experiment 2.3

Planning your own experiment

Aim

To show that chlorophyll is needed for photosynthesis.

You will need:
- a plant with variegated leaves
- equipment for a starch test

Method

1 Work in pairs. Discuss how you could use your equipment to show that plants need chlorophyll for photosynthesis.
2 Write down what you will do.
3 Carry out your experiment and record your results.

What can you conclude from your experiment?

Activity 2.3 Completing a table of results

Four de-starched green plants were kept in different conditions and then tested for starch.

1 Copy and complete the table to show what you think the results were.

Plant	Conditions the plant was kept in	Results of starch test
A	in sunlight with plenty of carbon dioxide	
B	in sunlight with no carbon dioxide	
C	in the dark with plenty of carbon dioxide	
D	in the dark with no carbon dioxide	

2 What do these results tell you about the things that plants need for photosynthesis?

Unit 4 Looking at leaves

Most photosynthesis takes place in the leaves of plants. Look carefully at Figure 2.8. Notice that the leaves:

- have a large surface area to absorb light
- are thin, so light and gases in the air can easily reach the cells inside
- are green, because they contain chlorophyll
- have veins, which transport substances to and from other parts of the plant.

↑ Figure 2.8 Leaves are well adapted for photosynthesis.

If you use a lens to look closely at the underside of a leaf, as in Figure 2.9, you can see tiny holes in the leaf surface. These are pores called **stomata**. Each stoma is surrounded by special cells called **guard cells** that allow the pore to open or close.

Stomata are important because:

- they allow carbon dioxide to diffuse into the plant
- they allow oxygen produced by photosynthesis to diffuse out of the plant
- they can open and close to stop the plant losing too much water.

➡ Figure 2.9
Stomata are small holes in the surface of a leaf.

The internal structure of a leaf

Figure 2.10 shows what a thin section of leaf looks like under a microscope. Read the information around the diagram carefully. It will help you to understand how well leaves are adapted for photosynthesis.

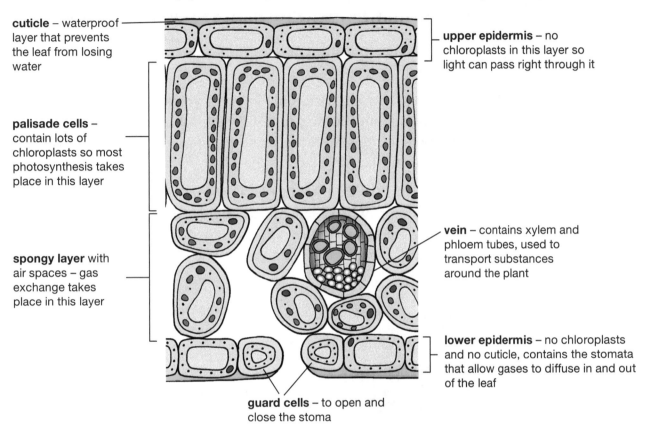

cuticle – waterproof layer that prevents the leaf from losing water

palisade cells – contain lots of chloroplasts so most photosynthesis takes place in this layer

spongy layer with air spaces – gas exchange takes place in this layer

upper epidermis – no chloroplasts in this layer so light can pass right through it

vein – contains xylem and phloem tubes, used to transport substances around the plant

lower epidermis – no chloroplasts and no cuticle, contains the stomata that allow gases to diffuse in and out of the leaf

guard cells – to open and close the stoma

↑ Figure 2.10 The internal structure of a leaf

Activity 2.4   Observing and reporting on leaves

Work with a partner. Collect three differently shaped and coloured leaves. Observe the upper and lower surfaces of each leaf through a lens.

1 Sketch the upper and lower surfaces of one leaf and label the parts you can identify.

2 Answer these questions about the leaves.
   a) Is the upper surface shinier or duller than the lower surface? Why do you think this is?
   b) Which surface is the darkest green? What makes it so green?
   c) Which surface gets the most light when the leaf is on the plant?
   d) Why are stomata found on the lower surface of the leaf?

Unit 5 How plants transport and store food

In Unit 4 you learned that most photosynthesis takes place in the leaves of a plant. So, plants make most of their food in their leaves. But the cells in every part of the plant need food to supply energy for life processes. This means that food has to be moved or **transported** around the plant.

Transporting food

Plants have a transport system made up of veins containing xylem (zy-lim) and phloem (flu-hm) tubes. Xylem tubes transport water from the roots to the rest of the plant. Phloem tubes transport food substances from place to place in the plant.

Glucose, which is made in photosynthesis, is water soluble – it dissolves in water. When plants need energy, dissolved glucose is moved along the veins (in the phloem tubes). The dissolved glucose can move to all parts of the plant to provide energy for life processes.

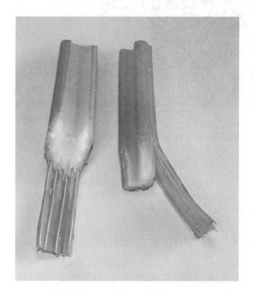

If the energy is not needed immediately, some of the glucose is converted into starch and stored in the plant. The starch can be changed back to glucose later if the plant needs the energy.

The rest of the glucose is converted into **cellulose**. The stringy fibres in the walls of plant cells are made of cellulose. Cellulose cannot be converted back into glucose if the plant needs energy.

→ Figure 2.11
The stringy fibres in these celery stems are made of cellulose.

Storing food

You have learned that plants use some of the glucose produced in photosynthesis straight away, and that some is changed into starch and cellulose and stored in the plant. Glucose can also be changed into other substances for storage.

Plants take up minerals such as nitrates and sulphates from the soil. Plants can combine these minerals with glucose to make amino acids. **Proteins** are built up from amino acids. Plants can also use minerals and glucose to form **lipids** (fats and oils).

Proteins and lipids are stored mostly in fruits and seeds. But plants can also store food in their roots, stems and leaves. Animals and humans eat these parts of plants to get the energy stored in them. Figure 2.12 shows some examples of how food is stored in different places in plants.

In carrots, food is stored in the roots.

Sugar cane plants store food in the stems.

The potato plant stores food in swollen underground stems.

The apple tree stores food in the fruits.

The leaves of spinach are rich in stored food.

Sunflower plants store most of their food in the seeds.

⬆ **Figure 2.12** Food is stored in different parts of plants.

Activity 2.5   **Describing how plants store food**

Look at the list of plants that humans eat. For each plant, write down which parts we eat. For example, the part of a banana plant that we eat is the fruit.

cabbage	onions	garlic	orange	beetroot	potato
tomato	radish	celery	okra	peanuts	sweet potato
lettuce	maize	peas	avocado	pumpkin	rice

Chapter summary

✓ Plants make their own food by the process of photosynthesis. We can represent photosynthesis using a word equation:

$$\text{carbon dioxide} + \text{water} \xrightarrow[\text{chlorophyll}]{\text{sunlight}} \text{glucose} + \text{oxygen}$$

✓ Glucose produced by photosynthesis is changed into starch for storage. When the plant needs energy, the starch is changed back to glucose and transported to the parts where it is needed.

✓ For photosynthesis to take place, the plant needs water, carbon dioxide and sunlight.

✓ Photosynthesis takes place mainly in the leaves of the plant. Leaves are well adapted for photosynthesis.

✓ The veins of plants contain phloem tubes, which carry food (dissolved glucose) around the plant to the places where it is needed.

✓ Plants store food in their leaves, stems, roots, fruits and seeds. Food can be stored as starch, converted to cellulose or used to make proteins and lipids.

Revision questions

1 Copy and complete these sentences.
  a) Green plants are able to make their own food by _____.
  b) A green pigment called _____ is found in the chloroplasts of plants.
  c) The raw materials for photosynthesis are _____ and _____.
  d) Glucose molecules produced by photosynthesis are made into larger _____ molecules for storage in the plant.
  e) _____ gas is a waste product of photosynthesis.

2 Give two examples of plant parts that swell up with stored food.

3 Write down the names of five plants that you eat regularly.
  a) Which part or parts of each plant do you eat?
  b) Say whether each plant is a good source of starch, protein or lipid.

Chapter 3 Plants from seeds

⬆ **Figure 3.1** These are seeds from different plants. Each one could grow into a new plant.

Last year, you learned that plants produce flowers, fruits and seeds. In this chapter, you will learn more about why plants do this and the role that flowers, fruits and seeds play in producing new plants.

As you work through this chapter, you will:

- learn how plants make new plants by producing flowers, fruits and seeds
- revise the parts of a flower
- define pollination and read about different kinds of pollen
- compare the ways in which different seeds are dispersed
- discuss the advantages of reproducing by producing seeds.

Unit 1 Making new plants

Plants and animals can grow, feed, respire and excrete for a certain time, but eventually all living things die. For life to continue, all living things need to reproduce themselves before they die. **Reproduction** is the process by which plants and animals form new living organisms.

Plants can reproduce themselves by flowering and producing seeds. The new young plants that grow from the seeds are similar to the parent plants, but not identical. Plants can also reproduce by growing a new plant from their roots or stems. When plants reproduce like this, the new plants are identical to the parent plant.

Seeds and variation

Producing seeds always involves two parent plants. Each parent makes **gametes**. A gamete is a special cell containing half of the genetic material that will form a new plant. So, when two gametes combine – one from each parent – the combined cell contains a full set of genetic material. This cell makes a seed.

Each gamete carries **characteristics** of the parent that made it. This means that when they combine, the seed they make will grow into a new plant with a mixture of characteristics – some from one parent and some from the other parent. This mixing means that new plants are like each of their parents in some ways, but different in others.

We say that reproduction by seeds introduces variety or **variation** within a species.

parent plants   offspring (new plants)

⬆ **Figure 3.2** All the flowers on the right are the offspring of the red and white parent plants. You can see that the new plants are similar to their parents, but not exactly the same.

To understand variation, think about your own family. You came from a male parent and a female parent. You probably look a bit like your mother and a bit like your father. However, you are not identical to either one of your parents because there is variation in your family.

Why is variation important?

Variation helps plants (and animals) to adapt to their environment and to survive in difficult conditions. **Adaptations** are features that help living things survive, especially in very hot, cold, dry or wet environments.

Variation means that some individual plants are able to grow or survive more successfully than others in their environment. For example, in a crowded habitat, taller plants will get more sunlight and therefore grow better than shorter plants. Some plants may be more resistant to disease or pests than others.

The offspring of strong, successful, disease-resistant plants are more likely to grow well than the offspring of weaker plants that are competing for the same resources.

Activity 3.1   **Talking about reproduction and variation**

1 Explain the meaning of each of these words or phrases.
   a) reproduction   b) gamete   c) characteristics   d) variation

2 Look at Figure 3.3.

→ Figure 3.3
Variation in cattle

   a) What examples of variation can you find between these members of the same species of cattle?
   b) What are the advantages for an animal like a cow of being bigger and stronger than other cows?

3 What do you think the phrase 'survival of the fittest' means?

Unit 2 The structure of flowers

Last year, you examined different kinds of flowers when you learned about flowering plants. There are millions of different kinds of flowers, each with their own colours, patterns, shapes, size and scent. However, all flowers have the same function: they contain the reproductive organs of the plant.

In a typical flower, there are four main parts: petals, sepals, stamen and carpel. The **stamen** is the male reproductive organ, consisting of anthers and filaments. The **carpel** is the female reproductive organ, consisting of the stigma, style and ovary.

Figure 3.4 shows the structure of a typical flower.

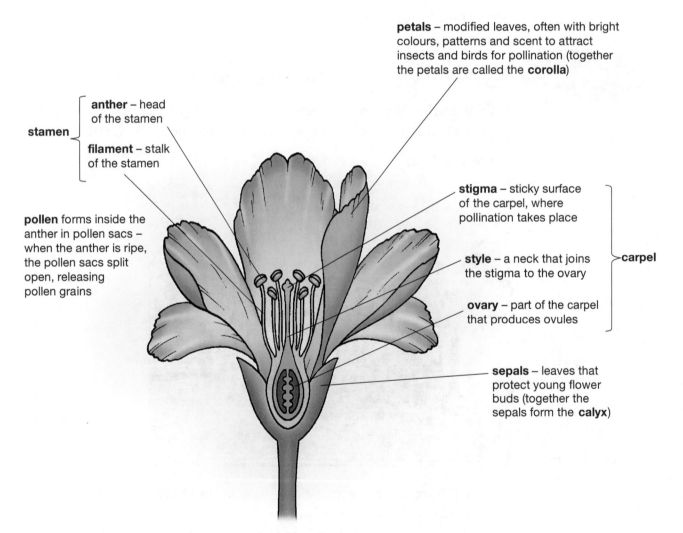

petals – modified leaves, often with bright colours, patterns and scent to attract insects and birds for pollination (together the petals are called the **corolla**)

anther – head of the stamen

stamen

filament – stalk of the stamen

pollen forms inside the anther in pollen sacs – when the anther is ripe, the pollen sacs split open, releasing pollen grains

stigma – sticky surface of the carpel, where pollination takes place

style – a neck that joins the stigma to the ovary

ovary – part of the carpel that produces ovules

carpel

sepals – leaves that protect young flower buds (together the sepals form the **calyx**)

↑ **Figure 3.4** The structure and parts of a typical flower

Experiment 3.1

Identifying and comparing parts of flowers

Aim
To examine flowers in order to identify and compare the parts.

You will need:
- three different kinds of flowers
- a hand lens

Method
1 Examine the flowers carefully, using a lens if you need to. Try to find the following parts of each flower:
- petals
- stamen (including anther and filament)
- carpel (including the stigma)
- sepals

2 Compare the parts of the different flowers. Use a table like this to record your observations.

Name of flower	Petals				Sepals				Stamens			Carpels	
	Number	Colour	Fused?	Separate?	Number	Colour	Fused?	Separate?	Number	Where attached	Longer/shorter than carpels?	Number	Longer/shorter than stamens?
A													
B													
C													

Activity 3.2 **Matching definitions to names**

Write down the correct name for each part being described.

1 This is brightly coloured to help attract insects to the plant.

2 This part contains the female reproductive organs.

3 This part contains the male reproductive organs.

4 This is a sticky surface that collects pollen.

5 This long stalk supports the anther.

Unit 3 Making seeds

→ **Figure 3.5** A seed inside an avocado

You already know that plants, like the avocado, make seeds. Two steps are needed for a seed to form: pollination and fertilisation.

Pollination

All flowering plants make **pollen**. Pollen is a fine powder produced by the anther. Most pollen is yellow, orange or brown in colour. You can see the pollen clearly on the anthers of the lily shown in Figure 3.6.

→ **Figure 3.6** These anthers are ripe – this means they are full of pollen.

For a seed to form, pollen has to travel from an anther to a stigma. When pollen lands on the stigma of a flower, the plant has been **pollinated**.

Pollen can be transported from the anther to the stigma by wind, water, insects, birds or other animals.

Plants that need animals to pollinate them are specially adapted to attract the right pollinator. Flowers with scent and colour help to attract pollinators. Plants that are pollinated by wind or water often have flowers with very small, dull petals or no petals at all because they do not need to attract pollinators.

The table summarises some of the different adaptations found in plants pollinated by insects or animals, and plants pollinated by wind or water.

Part of flower	Pollination by insects or animals	Pollination by wind or water
sepals	may be brightly coloured	often there are no sepals
petals	large, colourful and sometimes arranged in a pattern to attract particular insects or animals	small, green or brown, sometimes no petals at all
scent	may have a strong or sweet scent	no scent
stamens	may be hidden inside the flower, often sticky, with nectar to attract pollinators	long filaments with flexible, loose anthers that can blow in the wind
carpel	short, flat and sticky, usually enclosed inside the flower	exposed, may have long sticky extensions to catch pollen from the air
pollen	large, sticky, may be spiky	small, light

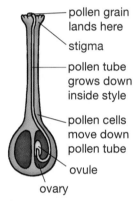

pollen grain lands here

stigma

pollen tube grows down inside style

pollen cells move down pollen tube

ovule

ovary

↑ **Figure 3.7** The pollen cells move down the pollen tube to reach the ovule.

Many plants cannot be pollinated by their own pollen. They need pollen from another plant (of the same type) to reach the stigma before pollination can take place. When one plant is pollinated by pollen from another plant, we say **cross-pollination** has taken place.

Fertilisation

When a pollen grain reaches the stigma, a small tube forms and grows down into the ovary. The pollen cells travel down this tube until they reach the ovary.

When a male pollen cell reaches the ovary, it joins with a female ovule to fertilise it. The new cell that is formed by **fertilisation** grows and divides to form a seed.

Activity 3.3 **Looking at adaptations for pollination**

ovule

↑ **Figure 3.8**

1 Study the flower in Figure 3.8 carefully.
   a) Where is the pollen produced on this flower?
   b) Is it easy to get to?
   c) Describe the stigma in this flower.
   d) Do you think this plant is pollinated by insects or by wind? Give a reason for your answer.

2 Why do plants pollinated by wind produce large amounts of very light pollen?

3 Why do some plants have sticky nectar in their flowers?

Unit 4 Seeds and fruits

Fertilised ovules develop into seeds. Each seed contains an **embryo** plant. The seed contains a store of food for the embryo plant. Most seeds have a hard or leathery coat to protect the food store and the embryo plant.

Once seeds are formed, the ovary starts to grow larger around them, to form the **fruit**. Fruits come in many different shapes and sizes, from huge juicy watermelons to tiny berries. Some fruits contain only one seed – for example, avocados, peaches, plums and mangos. On the other hand, fruits like the banana, fig, watermelon, papaya and orange contain many seeds.

The fruits help to protect developing seeds, and also help them to be **dispersed**, or carried away, from the parent plant before they start to grow.

Seed dispersal

Plants produce many seeds. This helps to make sure that at least some seeds will survive and grow into new plants. If a seed simply fell and landed right next to the parent plant, the new plant would have to compete with its parent for light, food and water. But if the seed is carried away, it has a better chance of landing somewhere where competition for resources is not so great. For this reason, seeds have many adaptations to help with dispersal. Figure 3.9 shows you some of these adaptations.

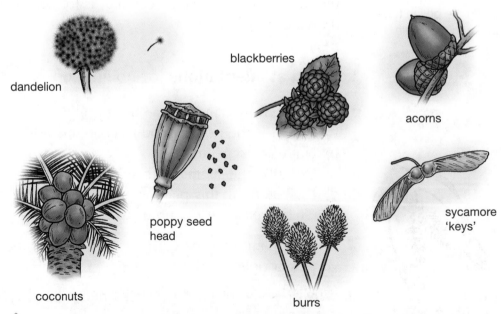

dandelion

blackberries

acorns

poppy seed head

coconuts

burrs

sycamore 'keys'

↑ **Figure 3.9** These plants are all adapted to help with seed dispersal.

Berries and nuts are eaten by animals. The enzymes in the animal's stomach often soften the outer coating of the seeds and this helps prepare the seed for **germination** (growing into a plant). The seeds pass through the animal's digestive system and pass out with the faeces. The faeces land on soil and the seed can begin to grow.

An acorn is a hardened ovary around a seed. Animals such as squirrels collect nuts like these and store them away from the parent plant. Some nuts drop in places where they can grow.

In some fruits, the ovary walls have tiny hooks on them. These hooks catch on the skin or fur of passing animals. When the animal moves around, the seeds can fall to the ground and grow.

Seeds that are very light are well adapted for dispersal by wind. Dandelion and grass seeds have feathery parachutes that help them float away from the parent plant. Others, such as sycamore seeds, have wings that help them to float on air. Other fruits, such as the coconut, are able to float so they can be dispersed by sea water and currents.

In some plants, like the poppy, the seed container dries out and grows hard. It pops open when the seeds are ready and scatters them away from the parent plant. This is called mechanical dispersion.

Activity 3.4   **Answering questions about seeds and fruits**

1 What do these terms mean?
  a) pollination
  b) fertilisation
  c) dispersal

2 Why is it best for seeds to travel away from the parents?

3 Why are some seeds small and light?

4 What types of seeds are eaten?

5 Why do some seed parts dry out?

6 How can grass seeds get from place to place?

7 How can a seed make a new plant when it has been eaten?

Chapter summary

✓ All living things reproduce. In flowering plants, reproduction involves two parent plants – one provides a male gamete and one provides a female gamete. A gamete is a special cell for reproduction.

✓ In a plant, the female gamete is in the ovule. The male gamete is in the pollen grain.

✓ Pollination takes place when pollen produced in the anthers is transferred to the stigma of a flower. The pollen might be transferred by an animal or insect, or by the wind.

✓ When a male pollen cell and a female ovule join, fertilisation takes place.

✓ The fertilised ovules develop into seeds, which contain embryo plants.

✓ Seeds develop inside fruits, which grow from the ovaries of the parent plant.

✓ Fruits and seeds are adapted to help with seed dispersal by wind, water, animals or mechanical means.

✓ Seed dispersal is important because seeds that germinate away from the parent plants do not compete with the parents for light, water and minerals, and so have a better chance of survival.

Revision questions

1 What has taken place when the pollen from one plant has reached a female reproductive organ in another plant?

2 What happens during fertilisation in flowering plants?

3 Why do flowering plants have seeds?

4 Do all plants make the same types of seeds? Give examples to support your answer.

5 Copy and complete this table to compare wind- and animal-pollinated flowers.

	Pollinated by animals	Pollinated by wind
Stigma		
Petals		
Nectar and scent		

6 If seeds dropped to the ground next to the parent plant, what problems would they face?

7 Why is it important that the seeds of large plants, like fruit trees, grow some distance away from the parent plant?

Adaptation and variation

↑ **Figure 4.1** Each of these zebra has its own pattern of stripes and no two patterns are identical.

You already know that plants and animals have adaptations that help them survive in the conditions around them. You also know that there are differences or variations between plants and animals of the same species, and that species vary from each other. In this chapter, you are going to revise what you already know about this topic. You will also find out more about variation and how new varieties of plants and animals come into being as a result of selective breeding.

As you work through this chapter, you will:

- revise how plants and animals adapt to conditions around them
- study examples of variation in plants and animals
- investigate variation in your environment
- understand what is meant by selective breeding
- identify new varieties of plants and animals produced by selective breeding.

Unit 1 Adaptations to different conditions

Plants and animals are adapted to the habitats they live in. An **adaptation** is a special characteristic that helps a plant or animal survive. Living organisms have adaptations that help them obtain food or water, move easily from place to place, protect themselves from predators, find shelter and reproduce.

Study the photographs and read the information to see how these plants and animals are adapted to living in different conditions.

⬆ Figure 4.2 This water lily is adapted to survive in water. The large flat leaves float on the surface, spread out in the light so that the plant can photosynthesise. The seeds float, so they can be dispersed by water.

⬆ Figure 4.3 Barnacles are adapted to survive in both wet and dry conditions. When the tide is out, they close up to protect themselves. When they are under water, they open up again and extend their tentacles to catch food.

➡ Figure 4.4 Mangrove trees are adapted to survive in river estuaries, where there is a mixture of fresh water and salty sea water, and where the water level changes with the tide. They have aerial roots with breathing holes above the water, and they excrete excess salt through special pores.

↑ Figure 4.5 Polar bears are adapted to live in the frozen Arctic. They have thick fur and blubber to keep them warm, and large feet to stop them sinking down into the snow. They are well camouflaged against the ice, and are very strong swimmers.

↑ Figure 4.6 These Andean people live in the high mountains of South America where the air contains lower levels of oxygen than air at sea level. Scientists have found that these people have more haemoglobin in their blood, which allows them to carry more oxygen with each breath.

Activity 4.1 Discussing adaptations

Work in small groups. Discuss each question with your group.

1 In some parts of the world, organisms can live and breed all year round because conditions do not change much. However, in some places, winters are very cold, the ground freezes and food is scarce. Discuss how animals and plants have adaptations to survive in places with very cold winters. Give examples.

2 How are cactuses adapted to living in hot, dry places?

3 Explain how fish are adapted to help them move in water.

4 How are birds adapted for movement?

5 Have humans adapted to living in different conditions? Give examples to support your answer.

Unit 2 Variation

Plants or animals belonging to the same species are similar in some ways. For example, all roses are similar in some ways. However, they also differ in some ways. For example, some roses are red, others are yellow, some are large and open, while some are smaller with tight petals. We can say that roses show **variation**.

In science, variation is the name given to the differences that exist between plants or animals in the same species.

The two animals shown in Figure 4.7 are both cats of the species *Felis catus*. They are alike in many ways because they share the characteristics of their species. For example, they are both four-legged mammals covered in hair, they have retractable claws and they miaow and purr. However, you can see clearly that the two cats are different from each other as well. For example, one is larger than the other, one has longer hair, and they are differently coloured. These two cats are a good example of variation between members of the same species.

➡ Figure 4.7 Two cats belonging to the species *Felis catus*

What causes variation?

When you studied reproduction in plants (page 20), you learned that young plants inherit some characteristics from each of their parents. These characteristics are combined in the young plant, and so it is similar to both parents but also different from them in some ways. Characteristics that are passed on from parents to offspring are called **inherited characteristics**. They are one cause of variation.

Some variation is caused by environmental conditions. For example, two plants of one species growing in different parts of a garden, where conditions are not the same, might look quite different from each other. One plant might be small, yellow and sickly because it does not have enough light and water – or perhaps it is exposed to pests or disease. The other plant might be tall, green and healthy because it has ideal growing conditions.

Environmental variation can be caused by:

- food – the amount and type of food animals eat, and the minerals available to plants, affect how they grow
- temperature – plants normally grow faster and better in warmer places
- physical training – when you use muscles their size and power increases, so athletes who train regularly look and perform differently to people who do not train
- diseases and injuries – poor environmental conditions allow the spread of preventable diseases such as malnutrition and tuberculosis, which can affect growth and development, and injuries can lead to physical differences or disabilities.

Some variation is caused by a mixture of inherited and environmental causes. For example, you might inherit good strong teeth from your parents, but if you eat lots of sweet foods and do not brush them properly you may grow to have weak teeth with lots of decay.

 **Activity 4.2**   **Tabulating causes of variation**

The box gives some examples of variations between humans. Copy the table and write each variation in the correct column.

Cause of variation		
Inherited	Environmental	Mixture of inherited and environmental

height	size of ears	ability to paint	shape of nose
ability to roll tongue	hair colour	length of hair	colour of skin
strength	mass	eye colour	size of feet
amount of tooth decay	language spoken	length of fingernails	acne

Unit 3 Selective breeding

Thousands of years ago, people noticed that young plants and animals are more similar to their parents than they are to other members of their species. So, people used this observation to breed plants and animals with useful characteristics.

For example, humans have bred cattle to produce more milk, fruit trees that bear more fruit and cereal crop plants that produce more grain. But how do they do this?

Let's say a farmer wants to breed sheep that produce more wool. The farmer notices that one of her sheep has very long and thick wool. She knows that the offspring of this sheep are likely to share this characteristic, so she decides to use this sheep for breeding. She mates the sheep with another animal that has this desirable characteristic. When lambs are born with long, thick wool, the farmer keeps these animals and uses them for breeding, when they are mature. Lambs without the long, thick wool are not used for breeding. Over many generations, more and more sheep are born with longer, thicker wool.

By choosing the characteristics she wants, the farmer has used **selective breeding** to produce sheep with long, thick wool. Picking out the best plants and animals for breeding is called **artificial selection**.

Producing new varieties

Another method of selective breeding involves producing new varieties of plants by cross-fertilisation, or cross-breeding. In this process, pollen from one variety of plant is used to fertilise another variety of the same type of plant. Study Figure 4.8 and read the information to see how selective breeding led to a new variety of rice.

→ **Figure 4.8** New varieties of plants with desirable characteristics can be produced by cross-breeding.

PETA is a tall rice variety from Indonesia. It produces lots of rice grains, but the tall plants are weak and cannot support the large heads of rice. The heads fall over into the water and the rice grains get ruined.

DGWG is a short, strong variety of rice from China. It is very strong but it doesn't produce large heads of grain.

IR-8 is a new variety of rice produced by cross-breeding PETA plants with DGWG plants. IR-8 has short, strong stalks, and is able to produce large heads of rice grains without falling over.

Advantages and disadvantages of selective breeding

Selective breeding has allowed humans to produce plants and animals with useful characteristics – such as crops that are disease resistant, and cattle that produce more milk. As a result, farmers have been able to produce more food.

The main disadvantage of selective breeding is that it reduces natural variation among the plants or animals. This could mean that large numbers of them would not survive if conditions changed.

Another disadvantage is that the characteristics the breeder wants may not be good for the organism. This is the case with some domestic animals that are bred for show, particularly cats and dogs. For example, Persian cats have been bred to have flat faces and long hair, because people find this appealing. But these characteristics mean that Persian cats often suffer from breathing problems, knotted hair and hairballs.

Activity 4.3 **Explaining selective breeding**

Today, domestic hens can lay around 250 eggs per year. But this was not always the case – long ago, hens each laid a few eggs per year, just like other birds. Today's hens are a result of selective breeding over many generations.

Explain how you would go about selectively breeding a hen that laid:
a) a large number of eggs   b) a smaller number of larger eggs.

Chapter summary

✓ An adaptation is a feature or behaviour that means a living organism is better able to survive in the conditions around it.

✓ Variation is the name given to the differences between members of the same species.

✓ Young plants and animals share more characteristics with their parents than they share with other non-related members of their species.

✓ Selective breeding involves choosing plants or animals that have characteristics we want, and breeding them. Some of their offspring are likely to inherit these desirable characteristics. The selection process is then repeated, choosing individuals from among the offspring to breed. Over generations, more and more offspring have the desirable characteristics.

Revision questions

1 How are the following organisms adapted to their natural habitats?
   a) barnacles   b) mangrove trees   c) polar bears   d) fish

2 Cats are mammals that are well adapted to living on land. Copy and complete this table to show how each adaptation is useful.

Adaptation	This feature is useful because . . .
body is covered in hair or fur	
babies develop inside the mother	
sensitive whiskers	
retractable claws	

3 Choose one person in your class who is not related to you in any way.
   a) Write down three differences that you can see between you and this person.
   b) Why do you have these different characteristics?
   c) What is the cause of the variation between you and this person?

4 A farmer raises cattle for beef. What characteristics should he look for in the cattle he is going to breed from?

5 A gardener wants to try and grow a giant pumpkin to enter into a competition. She chooses seeds from the largest pumpkin in her yard and plants them.
   a) Why did she choose seeds from the biggest pumpkin?
   b) What else does she need to do to make sure her pumpkin grows well?

Ecosystems and feeding relationships

⬆ Figure 5.1 Dung beetles play an important role in ecosystems.

Living organisms interact with each other and with the environments in which they live. In this chapter, you are going to learn more about the feeding relationships between organisms, using food chains and food webs. You will also learn about the factors that affect the sizes of different populations in ecosystems.

As you work through this chapter, you will:

- revise the vocabulary you need to talk about ecosystems
- understand how organisms depend on each other in an ecosystem
- classify organisms according to their role in feeding relationships
- work with food chains and food webs to explain and understand feeding relationships
- use diagrams to understand how energy flows through ecosystems
- work with ecological pyramids to describe feeding relationships
- identify factors that affect the sizes of populations.

Unit 1 Who eats who in an ecosystem?

Do you remember what we mean by an **ecosystem**? An ecosystem is a community of living organisms (plants and animals) and the physical conditions (soil, rocks, water) that they live in.

There are many different kinds of ecosystems, and each one has a set of conditions that makes it unique. Around the world, there are ecosystems in tropical and equatorial rainforests, mountain forests, grassland savannas, deciduous forests, scrubland, mangrove swamps, rivers, freshwater ponds, coral reefs and tidal pools. These ecosystems are all different from each other, but they all work in a similar way.

↑ **Figure 5.2** A tropical rainforest ecosystem in Australia

↑ **Figure 5.3** A semi-desert ecosystem in southern Africa

Producers and consumers

The Sun is the main source of energy for most ecosystems. The only way that living organisms can use energy from the Sun is by changing it from solar energy into chemical energy that they can use.

- You already know that plants can store light energy from the Sun as chemical energy in food (page 8). Some green algae and bacteria can also produce their own food. Organisms that are able to make, or produce, their own food are called **producers**.
- Animals cannot make food in their own bodies so they have to eat, or consume, other organisms in order to get food. Animals are called **consumers** in an ecosystem.
- Consumers that eat only plants are called **herbivores**. Sheep, rabbits, cows, elephants, giraffes and buffalo are all herbivores.
- Consumers that eat meat (in other words, they eat other consumers) are called **carnivores**. Lizards, seagulls, herons, owls, sharks, foxes and lions are all carnivores.
- Consumers that eat both plants and meat are called **omnivores**. Spider monkeys and humans are omnivores.

There are four feeding levels in most ecosystems, called **trophic levels**. You can see how these levels work in Figure 5.4.

trophic level 4 – the tertiary consumers, which eat secondary consumers

trophic level 3 – the secondary consumers, which eat primary consumers

trophic level 2 – the primary consumers, which eat the producers

trophic level 1 – the producers

leaves

locusts

sparrows

eagle

↑ Figure 5.4 **The different levels of feeding in an ecosystem**

↑ Figure 5.5 Decomposers have broken down the remains of this animal.

When organisms die, their bodies are broken down by a group of consumers called **decomposers**. Decomposers are bacteria and fungi that feed on dead and decaying material. As they feed, they release carbon dioxide, water and nutrients such as sulphates and nitrates into the soil. Decomposers play an important role because they make sure that nutrients stored in body tissues (food materials) are recycled and returned to the ecosystem.

Activity 5.1 Talking about ecosystems

ecosystem
producer
consumer
trophic level
decomposer

1 Make a table like this to summarise what you have learned about ecosystems. Include a row for each of the terms in the box.

Term	What it means	Local examples

2 Write down the names of ten animals that are found locally.
   a) Classify the animals as herbivores, carnivores or omnivores.
   b) Divide the carnivores and omnivores into secondary and tertiary consumers.

Unit 2 Food chains and food webs

Feeding transfers energy from the food to the feeder. When a primary consumer feeds, it gets energy from the plants it eats. The animal uses some of this energy in its own body, and loses some in the form of heat.

When a secondary consumer eats a primary consumer, it gets energy from the primary consumer. But the secondary consumer does not get all of the energy the primary consumer ate during its lifetime – the primary consumer used lots of that energy in its own body, and gave off lots in the form of heat, so there's not much left to be transferred. In fact, only about 10% of the energy from one trophic level is transferred to the next trophic level by feeding. This means that ecosystems can support only a limited number of tertiary consumers, because there's so little energy left at trophic level 4.

Food chains

We can show the feeding relationships and flow of energy between producers and consumers in an ecosystem using a simple diagram called a **food chain**. For example:

leaves → locust → lizard → owl

grass → caterpillar → bird

In general, food chains follow the order:

green plant → herbivore → carnivore

The arrows in a food chain show the direction in which energy moves, up the trophic levels.

Food webs

In most ecosystems, you can find different kinds of plants, so there is more than one primary producer. And most animals eat more than one thing. As a result, feeding relationships in most ecosystems are more complex than a food chain can show, so we use a **food web** instead. Figures 5.6 and 5.7 show simple food webs for two ecosystems.

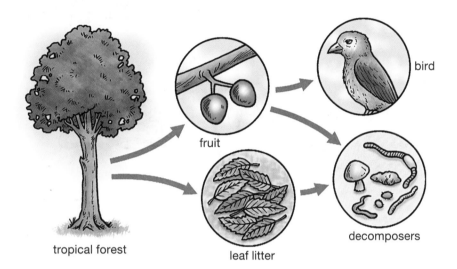

→ **Figure 5.6**
Some feeding relationships in a forest

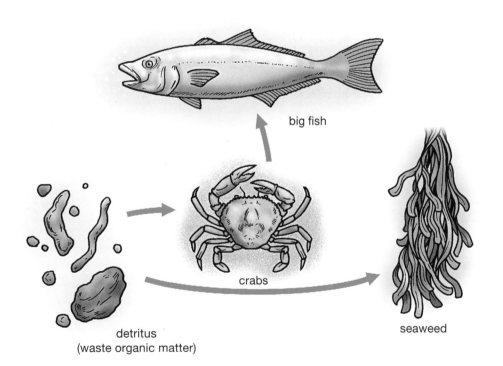

→ **Figure 5.7**
Some feeding relationships in a large rock-pool

A food web is really a combination of different food chains, showing how they are linked together. Figure 5.8 shows a food web for a coral reef. Figure 5.9 shows a food web for a cold-water ecosystem such as the Antarctic Ocean.

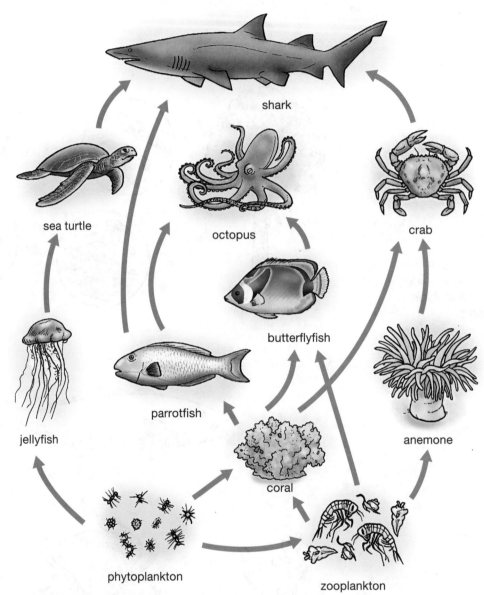

shark

sea turtle

octopus

crab

butterflyfish

anemone

jellyfish

parrotfish

coral

phytoplankton

zooplankton

⬆ **Figure 5.8** A coral reef food web

Remember that all food webs have the same characteristics:

● they all start with producers – this is the first trophic level
● they show consumers (herbivores, carnivores and omnivores), and the levels at which they feed
● the arrows show the direction in which energy flows – *from* the organism that is eaten *towards* the organism that eats it.

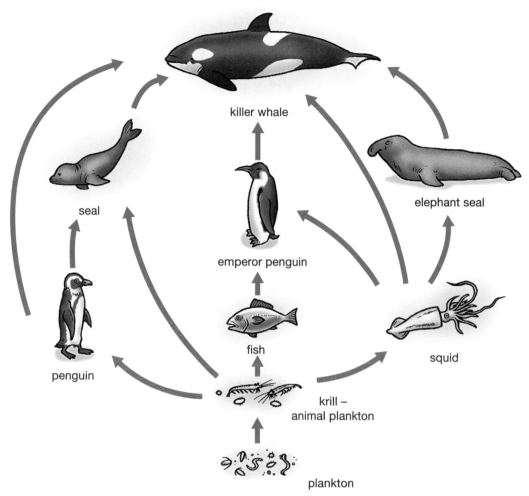

killer whale

seal

elephant seal

emperor penguin

penguin

fish

squid

krill –
animal plankton

plankton

⬆ Figure 5.9 An ocean food web

Activity 5.2   Interpreting a food web

1   Use the food web in Figure 5.9 to find the answers to
    these questions.
    a) Name the producers in this food web.
    b) Which are the primary consumers?
    c) Name a herbivore and two carnivores in this food web.
    d) What would be likely to happen to the number of seals in this
       ecosystem if the killer whale numbers were reduced? Why?
    e) If the krill died out because of disease, how would it affect
       the ecosystem?

2   Make up five questions of your own about the coral reef food
    web in Figure 5.8. Join with a partner and try to answer each
    other's questions.

Unit 3 Ecological pyramids

In Unit 2, we used food chains and food webs to show feeding relationships in ecosystems. However, these diagrams do not show the *numbers* of organisms at different trophic levels, nor do they show *how much* energy is transferred between the levels.

To show the numbers of organisms at different trophic levels, or the amount of energy transferred from level to level, scientists use a type of bar graph called an **ecological pyramid**.

Pyramid of numbers

A **pyramid of numbers** shows you how many organisms there are at each feeding level. For example, look at this food chain:

leaves → locust → sparrow → eagle

If we know how many leaves, locusts, sparrows and eagles there are, we can draw a pyramid of numbers. Figure 5.10 shows you what a pyramid of numbers would look like if there were 500 leaves, 100 locusts, 20 sparrows and 2 eagles.

Some pyramids of number appear to be upside down because one large plant is able to support many smaller consumers. Figure 5.11 shows you a pyramid of numbers for a large tree.

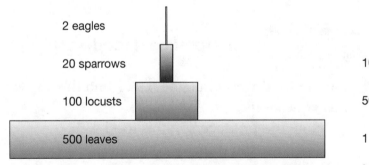

2 eagles

20 sparrows

100 locusts

500 leaves

↑ Figure 5.10 A pyramid of numbers shows how many organisms there are at each trophic level.

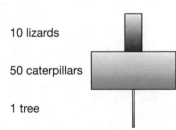

10 lizards

50 caterpillars

1 tree

↑ Figure 5.11 This is a pyramid of numbers for a food chain in a large tree.

Pyramids of biomass

Pyramids of biomass show the combined mass of all the organisms in a particular feeding relationship. Figure 5.12 shows a pyramid of biomass for the large tree shown in Figure 5.11.

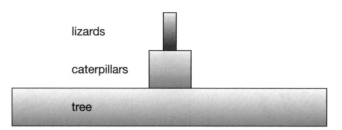

➔ Figure 5.12
A pyramid of biomass shows how much matter there is at each feeding level.

lizards

caterpillars

tree

Pyramids of energy

A **pyramid of energy** shows the amount of energy transferred from one trophic level to the next during a period of time, normally a year. Generally, these pyramids show that one-tenth of the energy from each level is transferred to the next. The smallest amount of energy is transferred to the top trophic level.

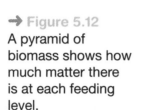

Activity 5.3   **Working with ecological pyramids**

1 Study the ecological pyramid for a freshwater lake in Figure 5.13 carefully.

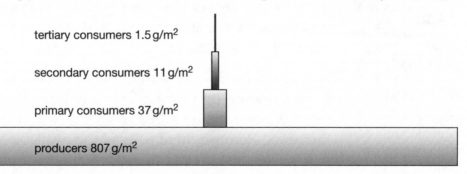

tertiary consumers 1.5 g/m^2

secondary consumers 11 g/m^2

primary consumers 37 g/m^2

producers 807 g/m^2

➔ Figure 5.13

   a) What type of pyramid is this?
   b) What are the producers likely to be in this ecosystem?
   c) Why do pyramids like this often have three or four levels, but seldom five?
   d) What are the tertiary consumers likely to be in this ecosystem?

2 A single rose bush supports a large community of aphids, which in turn provide food for ladybirds. Draw a rough sketch to show what the pyramid of numbers would look like for this food chain.

Unit 4 Population size

A **population** is a group of organisms of the same type that live in the same environment. For example, all the rhinoceros that live in a game reserve make up the population of rhinoceros in the reserve.

Factors that affect population growth

The size of a population depends on factors such as the amount of space and food available, disease, and the number of predators.

If more offspring are produced and survive, or if more organisms move in from another area, the population will grow.

If more organisms die because of predation, disease or lack of food, or if they go to live in another area, the population will decrease.

So, the size of a population, and the speed or rate at which it changes, depends on these four factors acting together:

- the birth rate
- the death rate
- movement into the population
- movement out of the population.

Imagine a population of mice living in a barn. The mice have a high birth rate. They also have a good supply of grain to eat and a sheltered place to live. Owls and other predators cannot reach them in the barn. As a result, death rates are low. In this example, the population of mice will grow because many mice are born and very few die.

Now, let's imagine the farmer brings traps into the barn, seals the grain in mouse-proof containers and locks three cats in the barn at night. Many mice are killed by the traps and cats, so the death rate increases. Because there is less food available, some mice leave the barn to live elsewhere. As a result, the barn population of mice declines, because the number of mice dying or leaving is greater than the number of mice being born.

population in barn

+

lots of offspring

−

a few deaths

=

population growth

↑ Figure 5.14
A high birth rate can lead to population growth.

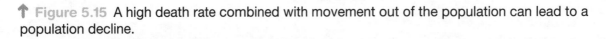

population in barn   few offspring   some mice move away   many deaths   population decline

↑ Figure 5.15 A high death rate combined with movement out of the population can lead to a population decline.

This table shows other factors that affect the rate at which populations grow.

Factor	Why it affects population size
how much food and water is available	If there is a drought in an area and the grass dies, the herbivore populations will probably decline, because less food is available to them. The herbivores may starve to death or move away.
predator–prey relationships	If the number of predators in an area increases, they will eat more of the prey species. This will increase the predator population even more (because they have more food) and the prey population will decline. As the prey population decreases, less food is available for the predators, and their numbers may decrease in turn.
competition	When plants or animals have to compete for the same resources, there may be less space, food and water available, and this in turn will affect the rate at which populations can grow.
disease and natural disasters	Outbreaks of disease can increase the death rate and cause a drop in population size. Natural disasters can kill organisms, as well as damaging the environment so populations die out or move away.

Activity 5.4   Explaining changes in a population

The graph shows the numbers of black rhinoceros found in Africa from 1970 to 2007.

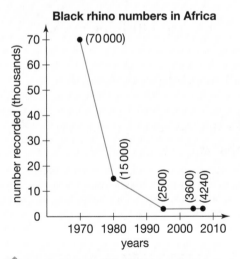

↑ Figure 5.16 Black rhinoceros numbers in Africa

1 Is the population of black rhino increasing or decreasing?

2 Suggest two factors that could have caused this population change.

3 Use information from the graph to complete this paragraph.

Populations of black rhino _____ very quickly during the last quarter of the twentieth century. In _____, it was estimated that 70 000 black rhino survived in Africa. By 1980, only around _____ remained and by 1995 the number had dropped to _____. The figure for 2007 indicates a total population of _____ black rhino. This shows that the population has _____ in recent times.

Chapter summary

ecosystems

producers and consumers
- plants
- herbivores
- carnivores
- omnivores

population change
- birth and death rates
- movement in or out of population
- food and water
- competition
- predator–prey relationships
- disease and natural disasters

feeding relationships
- food chains
- food webs
- ecological pyramids

Revision questions

1 Study the coral reef food web on page 42.
   a) Draw three food chains from this food web.
   b) Name the producers in the food web.
   c) What is the difference between a food web and a food chain?
   d) What would happen in this food web if the following consumers were removed: octopus, sea turtle, shark?
   e) What would happen in this food web if the number of parrotfish increased dramatically?

2 Draw a pyramid of numbers to show a feeding relationship in which there are 20 daisy plants, 10 caterpillars and 2 sparrows.

3 The table shows you some of the factors that help populations grow. Copy and complete the table by filling in the matching factors that cause populations to decline.

Factors that help populations grow	Factors causing population decline
plenty of space	
good food supply	
good water supply	
disease resistance	
favourable soil and climate (abiotic factors)	
lack of predators	

Chapter 6

Human activity and ecosystems

↑ Figure 6.1 Human activities, such as logging, can drastically change ecosystems.

'Human activity' means all the things that humans do as we go about our lives. This includes farming, mining, fishing, transport, making things and getting rid of wastes, as well as leisure activities. In this chapter, you are going to look at some of the ways in which human activities can affect natural ecosystems.

As you work through this chapter, you will:

- understand how ecosystems can change naturally, and as a result of people's actions
- describe some of the effects that human activities have on the environment
- do a survey to find out how your own environment has changed over time
- understand how pollution can affect ecosystems
- test a local water source to find out how clean the water is
- find out about biodiversity and why it is important in ecosystems
- identify ways in which humans can help to care for ecosystems and conserve biodiversity.

Unit 1 How do ecosystems change?

↑ Figure 6.2 This ecosystem changes all the time as the tide goes out and comes in.

Ecosystems change constantly. Some changes are slow and take place over a long time. For example, when a farmer stops ploughing a piece of land and lets it grow naturally, it takes time for the natural vegetation to regrow. Other changes are quick. For example, a storm or flood can change an ecosystem very quickly. Some changes do not last very long – we say they are short-term. For example, waves running over a beach cause short-term changes. Other changes are long-term or permanent, such as when a landslide changes the side of a mountain forever.

Some changes are a result of natural causes. For example, ecosystems change with the seasons. A wetland or swamp may dry up completely during the dry season. Trees may lose their leaves during colder months, and animals may migrate to other ecosystems at different times of year. Other natural changes are those caused by erosion or changing tides.

→ Figure 6.3 This wetland is flooded during the rainy season. It dries up completely in the summer months when there is no rain.

Micro-organisms can also cause changes in ecosystems. Micro-organisms are tiny living things too small for us to see, such as bacteria. Many decomposers are micro-organisms – we don't see them at work, but they break down dead and decaying matter, and change the ecosystem. The fungi on the tree trunks in Figure 6.4 are changing their ecosystem in this way.

← Figure 6.4 These fungi are changing the ecosystem, slowly and permanently.

Natural disasters can also cause massive changes to ecosystems. Floods, drought and even bush fires can result in long-term or permanent changes. When ecosystems change, organisms that lived there may no longer be able to survive. If an entire species dies out, we say the species has become **extinct**.

➜ Figure 6.5
The dodo and *Parasaurolophus* are examples of species that have become extinct.

dodo

Parasaurolophus

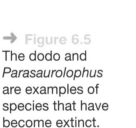

Activity 6.1   ## Discussing change

1 What types of natural events might cause living organisms to become extinct?

2 What sorts of human activities might cause living organisms to become extinct?

3 Does it matter if some plants and animals become extinct? Discuss this with your classmates and share your ideas with the rest of the class. Make sure you give reasons for your answers.

Unit 2 Human activities and ecosystems

In 2008, the United Nations estimated that there were 6.7 billion people living on Earth. People interact with the natural environment in many different ways. These interactions can change ecosystems, and make it difficult for the plants and animals living there to meet their needs.

The Worldwide Fund for Nature (WWF) estimates that, since 1960:

- the numbers of forest species have declined by 15%
- the numbers of marine (ocean) species have declined by 35%
- the numbers of freshwater species have declined by 54%.

In addition, scientists estimate that about 27000 species are becoming extinct every year – over 70 species per day! Most of these extinctions are caused by human activities that cause habitat loss.

Habitat loss

When the environment is changed, the organisms living there may lose their habitat. When people chop down forests, clear land for farming, dig mines or fill in wetlands, they are destroying the habitats of species that live there. These changes often cause the species to die out or move away. Then there are knock-on effects, because when one species is removed, the feeding relationships in an ecosystem change, often in negative ways.

Almost half the world's people live in cities. Building cities destroys natural habitats, and the noise and pollution of city living is unsuitable for most organisms that used to live in the environment.

Farming

In order to plant crops, people have to clear areas of natural vegetation and the ecosystem is completely changed. In most parts of the world, large areas of land have been changed by farming. Farmers also keep livestock, which eat natural vegetation and contribute to habitat loss.

← Figure 6.6 These rice paddies in Indonesia have completely replaced the tropical forest ecosystem that used to exist in this area.

Mining

Mining disturbs the environment, by removing vegetation and destroying land surfaces. It can also be very harmful because poisonous waste products sometimes seep into the ground, contaminating water supplies. From there, the poisons get into food chains.

→ Figure 6.7
Gold mining produces large waste dumps, which contain cyanide and other toxic chemicals. The water that runs out of them is highly poisonous.

Forestry

Cutting down trees destroys ecosystems, and plantations often replace natural vegetation and **indigenous** trees. Often the plantation trees (pines and gums), which are grown for wood and paper, do not support the indigenous species that used to live in the ecosystem.

Experiment 6.1

Investigating environmental changes

Aim

To find out how much your local environment has changed over time.

Work in pairs. Interview parents and grandparents to find out how your environment has been changed by human activity over the past 30 to 50 years.

Write a short report on your findings.

Activity 6.2 **Thinking about how humans affect ecosystems**

Write down three ways in which each of the following human activities negatively affects ecosystems.

1 building roads   2 farming   3 shopping   4 fishing

Unit 3 Pollution

One of the most damaging ways in which humans have affected ecosystems is through **pollution**. There are many different kinds of pollution. Fumes from cars and factories cause air pollution. Garbage and other waste pollutes the land. Noise is also a form of pollution. However, one of the most serious forms of contamination of an ecosystem is water pollution.

Water pollution

There are many different ways for water to become polluted. Factories may pipe polluted water directly into rivers, lakes or oceans. When it rains, the water washes pollutants from road surfaces and buildings into drains, and then it runs into the sea. Fertilisers and insecticides are washed off farmland and into rivers. Mud from construction sites can also flow into rivers and lakes and affect plant and animal life.

→ Figure 6.8
This canal has been seriously polluted by litter and other contaminants.

Pollutants act in various ways:

- they poison water supplies, leading to the death of plants and animals
- they may increase the temperature of the water, which in turn affects the plants and animals in the ecosystem
- they can reduce the amount of oxygen in the water, leading to the growth of algae and the process of **eutrophication**
- they encourage the growth of disease-causing organisms, which can have a serious impact on human health.

Experiment 6.2

Investigating water quality

Aim

To carry out some basic tests to determine the quality of a water source.

string black and white quarters marked with waterproof ink

Lower the disk into the water until you can't see it any more.

Measure the length of the wet part of the string.

↑ Figure 6.9
How to use a visibility disk

You will need:

- access to a local water source (stream, river, lake or dam)
- a visibility disk
- a tape measure
- a thermometer
- a glass jar with a lid
- a magnifying glass
- litmus paper
- an evaporating dish
- a Bunsen burner and tripod
- a funnel, conical flask and filter paper

Method

1 Draw a labelled sketch of the site where you will test the water source, showing the main features.
2 Use the disk to determine how clear the water is. Figure 6.9 shows how.
3 Use the thermometer to check the water temperature. Are there drains or pipes nearby that may feed warmer or colder water into the water source?
4 Describe the colour and smell of the water. Are there any oily 'rainbows' on it?
5 Collect some water in the jar. Shake it to see if bubbles form. What does this tell you?
6 Study your sample using the magnifying glass. Are there any living organisms in the water?
7 Use litmus paper to test the pH of the water.
8 Back in your classroom, evaporate a sample of the water in a dish over a Bunsen burner to see if there are any particles in it, or filter it through filter paper.
9 Write a short report on the general state of your water source. Give reasons for any statements you make.

Activity 6.3   Thinking about cause and effect

1 When water gets warmer, the amount of oxygen in it decreases.
   a) What could happen if a factory next to a river pumps thousands of litres of clean but warm water into the river?
   b) What can the factory do to lessen the environmental impact of the warm water?

2 Why is it a bad idea to wash your car next to the gutter or drain?

Unit 4 How can we prevent species loss?

Biodiversity is the great variety of living things and different ecosystems found on Earth. 'Bio' is short for biological, which means living, and 'diversity' means variety.

Why is biodiversity important?

You have seen that living organisms depend on each other to survive and that food webs are balanced. If the numbers of one species in a food web change significantly, it affects the whole ecosystem. Changes like this can damage or even cause the collapse of the ecosystem.

Can we do anything to conserve biodiversity?

Here is part of a news article about a recent United Nations report on loss of biodiversity.

Humans at the root of the problem

'The loss of biodiversity is a major barrier to development already and poses increasing risks for future generations,' said Walter Reid, the director of the Millennium Assessment. 'However, the report shows that the management tools, policies and technologies do exist to dramatically slow this loss.'

According to the report, changes in biodiversity due to human activities were more rapid in the past 50 years than at any time in human history, and over the last 100 years species extinction caused by humans has multiplied as much as 1000 times.

A proactive approach needed to solve problem

About 12% of birds, 23% of mammals, 25% of conifers and 32% of amphibians are threatened with extinction, and the world's fish stocks have been reduced by an astonishing 90% since the start of industrial fishing.

'We will need to make sure we don't disrupt the biological web to the point where collapse of the whole system becomes irreversible,' warns Anantha Kumar Duraiappah, of Canada's International Institute for Sustainable Development, one of the co-chairs of the report.

The report notes that, while efforts have helped reduce the loss of biodiversity, more action is needed as little progress is foreseen in the short term. The report blames biodiversity change on a number of factors including habitat conversion, climate change, pollution and over-exploitation of resources.

This list summarises some of the ways in which we can conserve ecosystems and preserve biodiversity.

- **Conservation areas** One method of protecting natural ecosystems is to fence them off as conservation areas. People may use these areas but may not disrupt the plants and animals.
- **Laws** Laws controlling pollution, preventing trade in endangered species and stating how land can be used, all help to protect biodiversity. For example, in some countries only unleaded petrol may be sold and used, to reduce air pollution.
- **Removal and control of alien species** In most countries, the customs services work to make sure that alien (non-indigenous) plants and animals are not imported into their country. This helps to prevent the spread of disease and also protects indigenous species by stopping alien plants and animals from taking over ecosystems.
- **Artificial or controlled breeding** Nurseries and seed propagation programmes help to conserve endangered plants. Endangered animals may be bred in captivity in zoos. In Africa, there are special programmes to try and raise the number of wild dogs, a highly endangered species.

↑ Figure 6.10 This game reserve is in southern Africa. Endangered ecosystems are protected here.

↑ Figure 6.11 Breeding programmes should help to increase the numbers of wild dogs.

Activity 6.4 **Taking personal action**

If we each take some action, however small, we can help in conserving biodiversity.

1 Work in groups. Think of some ways in which you could contribute to conserving or increasing biodiversity in your school or community.

2 Choose one project. Plan and carry out your choice in groups.

3 Decide how you will measure the success of your project. Monitor and evaluate it.

Chapter summary

✓ Ecosystems change all the time. Some changes are slow, while others happen quickly. Some changes are temporary and short-term, others are long-term or even permanent.

✓ Human activities such as farming, mining, forestry and building cities all contribute to changing ecosystems, often with negative effects on species that used to live there.

✓ Pollution of ecosystems is a serious problem. Water pollution is one of the most serious forms of pollution because it affects both plants and animals, and toxins in water can be transferred to food chains.

✓ Biodiversity is important because plant and animal species depend on each other for food and survival. The removal or destruction of one species can have a serious impact on others in a food web or ecosystem.

Revision questions

1 Give examples from this chapter or from your local environment to show what the following terms mean.
   a) natural changes
   b) micro-organisms
   c) extinct species
   d) human activities
   e) habitat loss
   f) indigenous vegetation
   g) pollution
   h) biodiversity
   i) conservation

2 Draw a mind map to summarise what you have learned in this chapter.

Chapter 7

Atoms and the Periodic Table

↑ **Figure 7.1** A random pile of beads can be sorted using the properties of the beads.

When the beads in Figure 7.1 were sorted and organised by colour, shape or size, it became much easier to tell the difference between them. This is similar to what scientists did by arranging the elements in the Periodic Table.

Last year, you learned about atoms and elements, and you worked with the Periodic Table to find and name the first 20 elements. This year, you are going to learn more about finding patterns among the elements, and understand how scientists used these patterns to draw up and add to the Periodic Table. You will also learn more about the organisation of the Periodic Table.

As you work through this chapter, you will:

- revise what you know about atoms and learn about electron configuration
- relate the structure of atoms to the positions of the elements in the Periodic Table
- describe the patterns found in groups and periods
- identify the names of important groups of elements
- distinguish between metals, non-metals and metalloids.

Unit 1 The structure of atoms

Do you remember the names of the parts of an atom?
Study Figure 7.2 carefully and make sure that you know that:

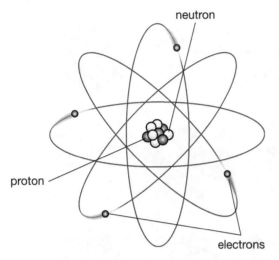

↑ **Figure 7.2 The parts of an atom**

- an atom is made up of many extremely small particles called sub-atomic particles
- the **nucleus** of an atom contains **protons** and **neutrons** – neutrons have no charge and protons have a positive charge
- **electrons** are smaller than protons and neutrons, and they carry a negative charge
- electrons move around the nucleus of the atom in orbits
- the number of protons in the nucleus of an atom is its **atomic number**
- in any atom, the number of protons is equal to the number of electrons – each positive charge (proton) is balanced out by a negative charge (electron), making the atom electrically neutral.

Electron arrangement

We don't know exactly how electrons move around the nucleus of an atom because we cannot see them. However, scientists have studied the behaviour of atoms and developed a theory, called the shell model, to explain how electrons are arranged.

The shell model

We can think of the electrons as arranged in groups, with each group orbiting at a different distance from the nucleus. The groups are called **shells** and we show them as a series of concentric circles. The first shell of an atom (the one which is closest to the nucleus) can hold up to two electrons. In other words, it can have one or two. The second and third shells can hold up to eight electrons each. The way in which electrons are arranged is called the **electron configuration** of the atom. Figure 7.3 shows the electron configurations of the first 20 elements.

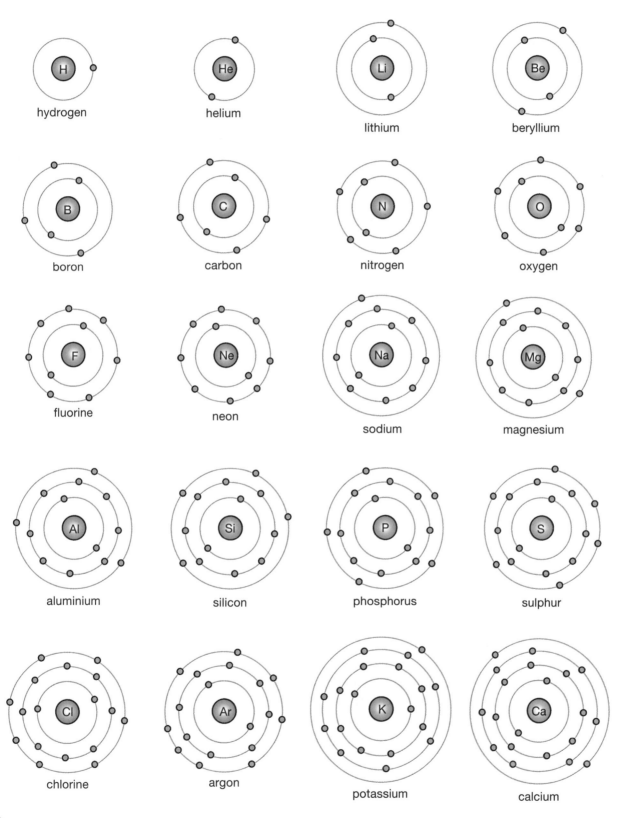

↑ **Figure 7.3** Electron configuration in shells of the first 20 elements in the Periodic Table

Working out electron configurations

The rules for filling electron shells are very simple:

- The shell closest to the nucleus must be filled (with two electrons) before electrons can go into the second shell.
- The second shell must be filled (with eight electrons) before electrons can go into the third shell.
- The third shell must be filled (with eight electrons) before electrons can go into the fourth shell.

You are only dealing with the first 20 elements, so none of the atoms will have more than four shells. However, most elements have atoms with more than four shells, and these are filled in different ways, depending on the position of the element in the Periodic Table. You will learn more about this if you study chemistry at higher levels.

The atomic number of an element (which you find on the Periodic Table) tells you how many protons are in each atom of this element. Since the number of electrons in the atom is equal to the number of protons, we can use the atomic number, together with the rules above, to work out the electron configuration of any atom (for the first 20 elements).

What is the electron configuration in an atom of sodium?

Sodium has an atomic number of 11.

So, each sodium atom must have 11 electrons.

2 electrons must go into the first shell.

That leaves 9 electrons.

8 of these will fill up the second shell.

That leaves 1 electron.

Therefore, the configuration of electrons in sodium is 2,8,1.

Core and valence electrons

Core electrons are electrons that are in full shells. If we use the example of sodium, the electrons in the first shell (2 electrons) and the electrons in the second shell (8 electrons) are both found in full shells. These are the **core electrons**.

The valence shell of an atom is the outermost shell that contains electrons. Electrons in the outermost shell are called **valence electrons**. In sodium, the outermost shell contains 1 valence electron.

When atoms bump into each other, only the valence shells meet. This means that the valence electrons are the ones most likely to react with other atoms. The number of valence electrons determines whether or not an atom will bond with other atoms. In other words, the valence electrons determine how reactive an element will be. You will learn more about reactivity later in this chapter.

Are all atoms of an element the same?

All atoms of an element have the same number of protons. But some atoms can have a different number of neutrons. When two atoms of the same element have different numbers of neutrons, we call them different **isotopes** of the element. Most elements have several isotopes.

Isotopes are named using their **mass number**. The mass number is the sum of protons and neutrons in the atom. For example, carbon-12 and carbon-14 are both isotopes of carbon. Their mass numbers are 12 and 14, respectively.

You can see that the name of an isotope allows us to work out how many neutrons its atoms have.

number of neutrons = mass number – atomic number

We know that carbon has an atomic number of 6 (all carbon atoms have 6 protons). So:

carbon-12 must have 6 neutrons (12 – 6 = 6)

carbon-14 must have 8 neutrons (14 – 6 = 8)

| **Activity 7.1** | **Working with electron shells** |

1 Study the diagrams showing the electron configurations of the first 20 elements, in Figure 7.3.
   a) Which elements have only one shell in each atom?
   b) Which element has the greatest number of electrons?
   c) Which elements have four shells?

2 Make a table to show the electron configurations of the five elements in the box and how many valence electrons they each have. The atomic numbers are shown in brackets.

helium He (2)
neon Ne (10)
silicon Si (14)
chlorine Cl (17)
vanadium V (23)

Unit 2 Understanding the Periodic Table

Dmitri Mendeleev was the first scientist to develop a version of the Periodic Table based on the properties of elements. He was so convinced that there were patterns in the elements that he left spaces for elements that were unknown at the time.

As scientists have discovered more elements and learned more about atoms, the Periodic Table has been refined and developed. Glenn Seaborg, an American scientist, reorganised the Periodic Table to include the new elements. It is his version that is largely used today.

← Figure 7.4 Dmitri Mendeleev is recognised as the 'father' of the Periodic Table.

← Figure 7.5 Glenn Seaborg discovered elements 93 to 102 and won the chemistry Nobel Prize in 1951.

| 1 | 2 | | | | | | | | | | | 3 | 4 | 5 | 6 | 7 | 8 |

| | | | | | | | | | | | | | | | | | H 1 hydrogen | | | | | | | | | He 2 helium |

Periodic Table:

1	2											3	4	5	6	7	8
Li 3 lithium	Be 4 beryllium											B 5 boron	C 6 carbon	N 7 nitrogen	O 8 oxygen	F 9 fluorine	Ne 10 neon
Na 11 sodium	Mg 12 magnesium											Al 13 aluminium	Si 14 silicon	P 15 phosphorus	S 16 sulphur	Cl 17 chlorine	Ar 18 argon
K 19 potassium	Ca 20 calcium	Sc 21 scandium	Ti 22 titanium	V 23 vanadium	Cr 24 chromium	Mn 25 manganese	Fe 26 iron	Co 27 cobalt	Ni 28 nickel	Cu 29 copper	Zn 30 zinc	Ga 31 gallium	Ge 32 germanium	As 33 arsenic	Se 34 selenium	Br 35 bromine	Kr 36 krypton
Rb 37 rubidium	Sr 38 strontium	Y 39 yttrium	Zr 40 zirconium	Nb 41 niobium	Mo 42 molybdenum	Tc 43 technetium	Ru 44 ruthenium	Rh 45 rhodium	Pd 46 palladium	Ag 47 silver	Cd 48 cadmium	In 49 indium	Sn 50 tin	Sb 51 antimony	Te 52 tellurium	I 53 iodine	Xe 54 xenon
Cs 55 caesium	Ba 56 barium	La 57 lanthanum	Hf 72 hafnium	Ta 73 tantalum	W 74 tungsten	Re 75 rhenium	Os 76 osmium	Ir 77 iridium	Pt 78 platinum	Au 79 gold	Hg 80 mercury	Tl 81 thallium	Pb 82 lead	Bi 83 bismuth	Po 84 polonium	At 85 astatine	Rn 86 radon

key

| element symbol   atomic number |
| element name |

☐ non-metals   ☐ metals   ☐ alkali metals   ☐ alkaline earth metals

☐ transition metals   ☐ metalloids   ☐ halogens   ☐ inert gases

↑ Figure 7.6 Part of the modern Periodic Table. There are more elements (not shown) between lanthanum and hafnium and beyond radon, but they are all rare.

You worked with the modern Periodic Table last year. You should remember that it is like a map of all the known elements, arranged in rows and columns in order of atomic number. The atomic number is given above the symbol for each element (at the top inside each box). Each element's full name is also given in Figure 7.6 – some versions include only the symbol, and provide other information about the atoms that scientists need.

Groups

The columns in the Periodic Table are called **groups**. In some versions, every column is numbered to give 18 groups, but other versions only number eight groups. These are shown above the columns (sometimes in roman numerals). You only need to know the names of four main groups.

Notice that hydrogen stands alone. It doesn't fit into any of the eight groups. This is because atoms of hydrogen are unique – they do not have any neutrons, just one proton and one electron.

The elements in a group share some chemical properties. We can say that they react and change in the same way under certain conditions.

- **Group 1 – the alkali metals** The elements in this group are called alkali metals because they react with water to form an alkali. (Remember that an alkali is a substance that forms a chemical salt when combined with an acid.) If you look at the shell diagrams for elements in this group (Figure 7.3) you will see that they each have only 1 valence electron in their outer shell. This makes them highly reactive.
- **Group 2 – the alkaline earth metals** The alkaline earth metals give oxides and hydroxides that dissolve slightly in water to make alkaline solutions. Most of the medicines that people take for heartburn or gastric upsets contain compounds of calcium and magnesium.
- **Group 7 – the halogens** The name halogen means 'salt former'. The atoms of elements in this group all have 7 valence electrons and this makes them highly reactive.
- **Group 8 – the noble or inert gases** The elements in this group are all gases that are very stable and unreactive. Their atoms all have a full valence shell and this stops them from bonding easily with other elements.

Notice that the group number (from 1 to 8) tells you how many valence electrons there are in atoms of that group.

Changes as you go down a group

The physical properties of elements change as you go down a group. In general, melting points and boiling points decrease, while density usually increases as you go down the group. There are exceptions, though, and not all elements fit the pattern exactly.

The tables show you some of the physical properties for elements in Group 1 and Group 8.

Group 1

Element	Density (g/cm^3)	Melting point (°C)	Boiling point (°C)
lithium	0.534	180.7	1342
sodium	0.971	98	883
potassium	0.862	63.4	759
rubidium	1.63	39.6	688

Group 8

Element	Density (g/litre)	Melting point (°C)	Boiling point (°C)
helium	0.18	−272.1	−268.8
neon	0.9	−248.4	−245.9
argon	1.78	−189.2	−185.7

As you move down the elements in a group, the speed at which they react with other elements changes.

In general:

- in the groups of metals on the left-hand side of the Periodic Table, the elements become more reactive as you move down the group
- in the right-hand groups – the halogens and the noble gases – the elements become less reactive as you move down the group.

There are exceptions to these trends. For example, nitrogen, which is at the top of group 5, is not very reactive at all.

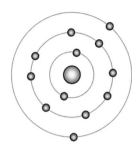

Periods

The rows across the Periodic Table are called **periods**. The elements in each period do not share the same physical properties but they do have same number of electron shells. Elements in the first period have only one electron shell, those in the second period have two, those in the third period have three, and so on.

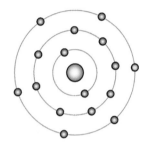

The first element in each period is usually a reactive metal, while the last element is a noble gas. As you go across the period from left to right, there is a gradual change from elements that are metals to elements that are non-metals.

The metalloids

You already know that the elements can be divided into two large groups: the metals and the non-metals. Some elements don't fit nicely into either of these groups. This gives us a third group called **metalloids**.

If you look again at the Periodic Table in Figure 7.6, you will see that there is a thick line stepping down between the elements from boron to polonium. This line separates the metals from the non-metals. The elements on either side of this line (shaded dark blue in Figure 7.6) are the metalloids, which have some properties of metals but not all. Silicon is probably the best-known metalloid. It is a shiny solid, which is very brittle. Silicon chips are used in computers and other technologies.

↑ **Figure 7.7**

Activity 7.2 **Applying what you have learned**

1 How are the elements in the Periodic Table arranged? Explain simply in your own words.

2 Which group of metals is most reactive?

3 Which group of non-metals is most reactive?

4 Besides being gases at room temperature, what other property do the noble gases share?

5 Study the three diagrams of atoms in Figure 7.7.
   a) Identify each of these elements.
   b) How many valence electrons does each atom have?
   c) To which group in the Periodic Table does each of these elements belong?

6 Choose one element in the Periodic Table that you would expect to share properties with:
   a) argon   b) sodium   c) fluorine   d) calcium.

Chapter summary

✓ The nucleus of an atom contains protons and neutrons.

✓ Protons have a positive charge, neutrons have no charge.

✓ Electrons have a negative charge.

✓ The number of protons in an atom is equal to the number of electrons. The positive and negative charges cancel out, making atoms electrically neutral.

✓ The nucleus is surrounded by electron shells. The way electrons are arranged in shells is called the electron configuration.

✓ Full shells contain core electrons. The outermost shell of electrons is called the valence shell and it contains the valence electrons.

✓ The number of valence electrons in an atom determines how reactive the element is.

✓ Isotopes of an element have the same number of protons but differing numbers of neutrons.

✓ In the Periodic Table, elements are arranged in columns called groups and rows called periods.

✓ Group 1 contains the reactive alkali metals, Group 2 contains the alkaline earth metals, Group 7 contains the very reactive halogens and Group 8 contains the very unreactive noble or inert gases.

Revision questions

1 Draw and label a diagram to show the meaning of the following terms:

- nucleus
- proton
- neutron
- electron
- core electrons
- shell
- valence shell
- valence electrons

2 Copy Figure 7.8. Add labels and notes to explain how the Periodic Table is organised and how it helps us see patterns in the elements.

➜ Figure 7.8

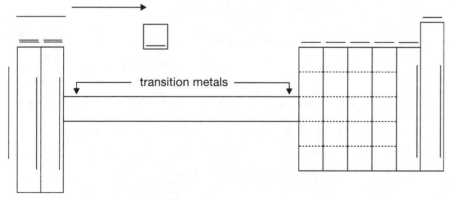

transition metals

Chapter 8

Reactions of metals and their compounds

↑ **Figure 8.1** Solids like these crystals are chemical compounds in which atoms have chemically bonded together.

You already know that atoms of different substances can combine chemically to form compounds. When substances combine chemically, we say that a chemical reaction has taken place. Last year you learned about different kinds of chemical reactions and their results. This year you are going to learn more about reactions of metals and metal compounds, and carry out experiments of your own to prepare salts.

As you work through this chapter, you will:

● revise different types of chemical reactions
● understand the difference between a molecule and an ion
● investigate what happens when metals react with dilute acids
● apply what you already know to test for gases produced in reactions
● investigate what happens when metal carbonates react with acids
● find out how to balance a chemical equation to represent a reaction.

Unit 1 Different types of chemical reactions

Do you remember the difference between a physical change and a chemical change?

↑ Figure 8.2
Melting is a physical change. Burning is a chemical change.

When a **physical change** happens, the substance might look different. For example, when ice melts it forms water. But the substance is not changed chemically. If you freeze the water, it turns back to ice. Physical changes are normally easy to reverse.

In a **chemical change**, new substances are formed. For example, when a camp fire burns, you are left with black carbon and ash, not wood. Chemical changes are not easy to reverse. A chemical change is also called a **chemical reaction**.

Look out for one or more of these five signs to tell you whether a chemical reaction is taking place:

- a gas is produced
- an insoluble substance (a precipitate) is formed
- there is a permanent change in colour
- there is a change in mass
- there is heating or cooling (a temperature change).

Most chemical reactions can be divided into one of four types:

- synthesis reactions (including combustion and combination)
- decomposition reactions
- single displacement reactions
- double displacement reactions.

The table gives examples of each type of reaction and summarises what happens.

Type of reaction	What happens	Examples
synthesis reactions (also called combination reactions)	two substances react or combine to form a new substance	• hydrogen gas (H_2) and oxygen gas (O_2) react to form water (H_2O) • when coal burns, solid carbon (C) reacts with oxygen gas (O_2) to form carbon dioxide gas (CO_2)
decomposition reactions	a single substance breaks down, or decomposes, to form simpler compounds or elements	• water can be decomposed (using electricity) to form hydrogen gas (H_2) and oxygen gas (O_2) • if you heat hydrogen peroxide (H_2O_2) it decomposes into water (H_2O) and oxygen (O_2)
single displacement reactions	one element or compound is moved and joined onto another	• when zinc (Zn) reacts with copper sulphate ($CuSO_4$) we get zinc sulphate ($ZnSO_4$) and copper (Cu) – zinc has displaced copper in the compound
double displacement reactions	two compounds react to produce two new compounds	• silver nitrate ($AgNO_3$) reacts with sodium chloride (NaCl) to form silver chloride (AgCl) and sodium nitrate ($NaNO_3$) – the chloride and nitrate groups have changed places

Activity 8.1   Classifying reactions

1 Which of the following are chemical changes?
   a) water boiling
   b) iron rusting
   c) decomposition of animal remains by bacteria and fungi
   d) a puddle evaporating

2 What type of reaction is represented by each of these?
   a) carbon + oxygen → carbon dioxide
   b) water → hydrogen + oxygen
   c) zinc + copper sulphate → copper + zinc sulphate
   d) silver nitrate + sodium chloride → silver chloride + sodium nitrate
   e) copper sulphate → copper oxide + sulphur dioxide
   f) magnesium + hydrogen chloride → magnesium chloride + hydrogen
   g) sodium oxide + hydrogen oxide (water) → sodium hydroxide
   h) hydrogen chloride + sodium hydroxide → sodium chloride + water

Unit 2 Atoms, ions and molecules

To understand how chemical reactions happen, and why atoms stick together to form matter, you need to understand a little more about particles and how they behave.

All matter is made of three types of particles: atoms, ions or molecules.

An **atom** is the smallest particle of an element. Atoms have a nucleus containing protons and neutrons, and electrons that move around the nucleus in a series of shells.

Atoms combine or **bond** with other atoms in different ways. The type of bond an atom makes with another atom depends on how many valence electrons there are in their outer shells. In simple terms, atoms bond to try to get a full outer shell – atoms may share, lose or gain electrons to get a full outer shell.

Covalent bonding

When the atoms of non-metal elements or non-metal compounds bond, they tend to *share* electrons. The two bonding atoms each provide an equal number of electrons, which form pairs that orbit the nuclei of both atoms. This type of bonding is called **covalent bonding** and it holds the atoms together as a **molecule**.

→ Figure 8.3
Models showing atoms joined together to form different molecules.

We can use models like those shown in Figure 8.3 to represent molecules because each molecule of a substance has a fixed number of atoms. For example, a molecule of hydrogen gas always consists of two atoms of hydrogen joined together. In the same way, a molecule of water is always made of two atoms of hydrogen combined with one atom of oxygen. The **chemical formula** of a molecule tells us how many atoms of each element are found in each molecule of the substance.

There are many different kinds of molecules. Water, oxygen, carbon dioxide, methane, plastics, proteins and human DNA are all examples of molecules.

How molecules fit together

Scientists observe how molecules behave when they mix together with other molecules, and make conclusions about their shape and size. You can do Experiment 8.1 to help you understand how the shapes and sizes of molecules affect what happens when they are put together.

Experiment 8.1

Investigating molecule size

Aim
To find out how the size and shape of molecules can affect what happens when they are put together.

You will need:
- two 100 cm³ measuring cylinders
- water
- ethanol

Method
1 Measure exactly 50 cm³ of water in one measuring cylinder.
2 Measure exactly 50 cm³ of ethanol in the other measuring cylinder.
3 Carefully pour the ethanol into the water.

Questions
1 How much liquid do you have in the measuring cylinder after combining the water and the ethanol?
2 Did you observe any changes when you combined the two substances? For example, did the cylinder get hotter or colder, did any bubbles form, was a gas given off?
3 Try to explain why 50 cm³ of water + 50 cm³ of ethanol doesn't give you 100 cm³ of liquid.

↓ **Figure 8.4** A mixture of large and small beads takes up less space than the two original amounts of beads.

You can understand what happened in Experiment 8.1 by combining two sets of beads. Look at Figure 8.4 and read the information carefully.

This beaker contains large beads up to the 100 cm³ mark.

This beaker contains small beads up to the 100 cm³ mark.

This beaker contains 100 cm³ of large beads mixed with 100 cm³ of small beads. Together the beads only come up to 180 cm³.

The bead experiment can help us to understand what happens when molecules are combined. When large beads are mixed with small beads, the smaller ones fill the spaces between the larger ones and the end result takes up less space than the two original substances did.

In the same way, if you mix molecules of different shapes, they may slot closely between one another and take up less space together than they did separately.

Metallic bonding

In metals, the atoms also share electrons to make **metallic bonds**. In this type of bonding, the electrons form a common supply that bonds all the metal atoms strongly together. You will learn more about metallic bonding at higher levels of chemistry.

Ions and ionic bonding

Remember that atoms are electrically neutral because they have an equal number of positive protons and negative electrons.

When atoms of a metal react with atoms of a non-metal, they do not share electrons. Instead, these atoms *gain* or *lose* valence electrons.

- An atom that loses a valence electron will have more protons (+) than electrons (−), so it will become positively charged.
- An atom that gains a valence electron will have more electrons (−) than protons (+), so it will become negatively charged.

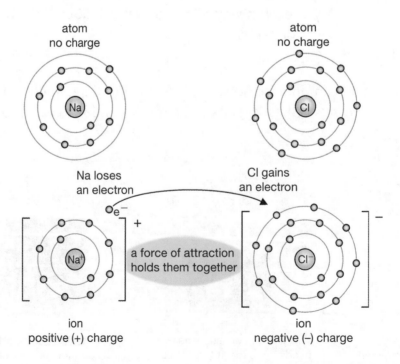

→ **Figure 8.5**
Sodium chloride is an ionic compound.

When atoms gain or lose electrons and become charged, they are called **ions**.

The positive ion and the negative ion have opposite charges, so they attract each other. This force of attraction is called **ionic bonding**. Ionic compounds form strong lattice structures a bit like the framework of a high building. The ions are tightly held together because the opposite charges attract each other strongly.

Figure 8.5 shows how sodium atoms react with chlorine atoms to form ions.

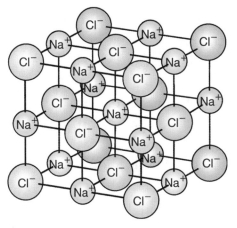

↑ **Figure 8.6** The particles in common salt (NaCl) are called ions, not molecules.

Figure 8.6 shows how the sodium and chloride ions bond together in a lattice structure, to make the ionic compound called sodium chloride (NaCl). Another name for sodium chloride is common salt.

Small negative and positive signs are used with the chemical symbol for the element, to show that atoms have become ions. So Na^+ represents a positively charged sodium ion, and Cl^- represents a negatively charged chloride ion.

In a molecular compound, the chemical formula tells you how many of each type of atom there are in one molecule. The formula of an ionic compound gives the ratio of the different types of ions present in the compound. In NaCl there is one sodium ion for every chloride ion, a ratio of 1:1.

Activity 8.2 Applying what you have learned

1 What is a molecule? Give two examples of molecules containing only one element.
2 What part of an atom is involved in atomic bonding?
3 What is a covalent bond?
4 What happens to the valence electrons of each atom when sodium is heated in chlorine gas?
5 Name the two ions found in sodium chloride (salt).
6 What is ionic bonding?
7 An atom of sodium has 11 protons and 11 electrons. How is this different from an ion of sodium?
8 Why do metals tend to form positive ions while non-metals form negative ions? Draw a diagram to support your answer.

Unit 3 Chemical salts

What do you think about when you hear the word 'salt'? Most people think of the white crystals we sometimes put on our food, or use in cooking – ordinary table salt. But, in science, we use the word **salt** differently.

Chemical salts

In science, salts are crystalline compounds. Salts are formed when metal or metal compounds react with acids. There are many different kinds of salts. Here are some examples of salts and their uses in daily life:

- sodium bicarbonate (baking soda) – used in cooking and baking and in medicines to neutralise stomach acid
- calcium carbonate – used to make classroom chalk as well as cement and other building materials
- potassium nitrate – used in fertilisers and to preserve meats
- aluminium phosphate – used in some medicines to absorb toxic substances
- magnesium sulphate (Epsom salts) – used in medications as a laxative.

If you look at the names of the salts, you will see that they contain the name of the metal (sodium, calcium, potassium, aluminium, magnesium) plus the name of part of the acid used in the reaction in which they formed. The type of salt formed depends on the type of acid used in the reaction. The table shows the salts formed by three acids you might use in school. Other groups from acids include carbonates (CO_3) and phosphates (PO_4).

Acid used	Salt formed
nitric acid (HNO_3)	nitrate (NO_3)
sulphuric acid (H_2SO_4)	sulphate (SO_4)
hydrochloric acid (HCl)	chloride (Cl)

Preparing salts

When you react a metal with a dilute acid you get a salt, and the reaction produces hydrogen gas. (Do you remember how to test for hydrogen gas? If not, read through page 4 again.)

Reacting a metal with a dilute acid

Aim

To observe what happens when you react magnesium ribbon with dilute hydrochloric acid.

You will need:
- two test tubes • a collecting tube • magnesium ribbon
- dilute hydrochloric acid • dilute sulphuric acid
- a lighted splint to test for hydrogen gas

Method

1 Put a small amount of dilute hydrochloric into a test tube.
2 Add a small piece of magnesium ribbon to the acid. Observe what happens.
3 Collect the gas that escapes, using the collecting tube. Figure 8.7 shows how.
4 Test the gas with a lighted splint. Observe what happens.
5 Repeat the experiment using dilute sulphuric acid.
6 Record the results of both experiments.

Questions

1 What do you see when you put magnesium ribbon into dilute acid?
2 How do you know that the gas is hydrogen?
3 What is left in the test tube?
4 Predict what would happen if you repeated this experiment using small shavings of zinc instead of magnesium ribbon.

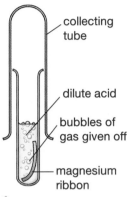

collecting tube

dilute acid

bubbles of gas given off

magnesium ribbon

↑ Figure 8.7
How to collect the gas from your reaction

You should remember, from work you have done before on acids and bases, that all acids contain hydrogen. When an acid reacts with a metal, the hydrogen is given off in the form of hydrogen gas (H_2). The rest of the particles of acid remain behind in solution with the metal. The solution is a salt. In Experiment 8.2, you prepared two salts:

- magnesium chloride ($MgCl_2$)   • magnesium sulphate ($MgSO_4$)

If you had repeated the experiment with zinc shavings, you would have produced these salts:

- zinc chloride ($ZnCl_2$)   • zinc sulphate ($ZnSO_4$)

We can represent what happens in these reactions in a word equation:

$$acid + metal \rightarrow salt + hydrogen$$

Making salts from metal compounds and acids

You know that metals can combine with oxygen to form oxides, and with carbon to form carbonates. These metal compounds react with acids to form salts, but the products of the two reactions are slightly different.

Metal oxides plus acid

Most metal oxides (and hydroxides) are bases. Remember that a base is the opposite of an acid. When acids and bases are mixed together, they neutralise each other, so we call the reaction of a metal oxide with an acid a **neutralisation reaction**.

Study Figure 8.8 and read the information carefully to learn what happens when copper oxide reacts with dilute sulphuric acid.

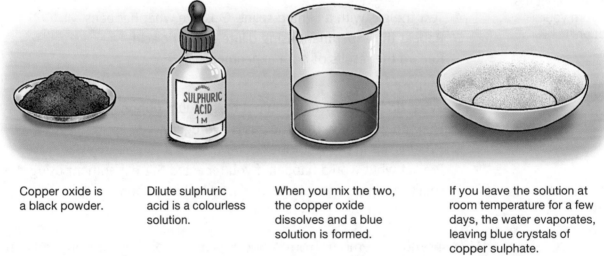

Copper oxide is a black powder.

Dilute sulphuric acid is a colourless solution.

When you mix the two, the copper oxide dissolves and a blue solution is formed.

If you leave the solution at room temperature for a few days, the water evaporates, leaving blue crystals of copper sulphate.

⬆ Figure 8.8 The reaction of a metal oxide and an acid produces a salt and water.

We can represent what happens in this reaction in a word equation:

copper oxide + sulphuric acid → copper sulphate + water

Metal carbonates plus acid

You can do experiments in the laboratory to see what happens when a metal carbonate, such as copper carbonate or calcium carbonate, reacts with a dilute acid. Study Figure 8.9 carefully to learn what happens when you react calcium carbonate with dilute sulphuric acid.

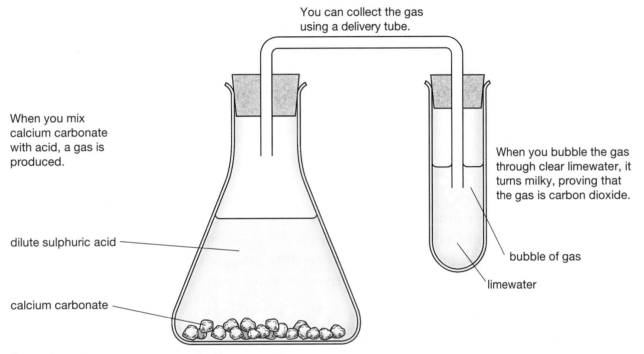

You can collect the gas using a delivery tube.

When you mix calcium carbonate with acid, a gas is produced.

When you bubble the gas through clear limewater, it turns milky, proving that the gas is carbon dioxide.

dilute sulphuric acid

calcium carbonate

bubble of gas

limewater

↑ **Figure 8.9** The reaction of a metal carbonate and an acid produces a salt, carbon dioxide and water.

We can represent what happens in this reaction in a word equation:

$$\text{calcium carbonate} + \text{sulphuric acid} \rightarrow \text{calcium sulphate} + \text{carbon dioxide} + \text{water}$$

Activity 8.3   Writing general equations for reactions

1 Complete these word equations to summarise what happens when a metal or a metal compound reacts with an acid.
   a) metal + acid →   b) metal oxide + acid →   c) metal carbonate + acid →

2 A metal is reacted with each of the following acids. What is the second part of the name of the salt you world expect to be produced in each case?
   a) nitric acid   b) hydrochloric acid   c) sulphuric acid.

Unit 4 Balancing chemical equations

Let's look at the reaction of hydrogen and oxygen, making water, to see how to write a chemical equation. We know that:

$$hydrogen + oxygen \rightarrow water$$

The **reactants** in this process are hydrogen and oxygen. The **product** is water (H_2O).

Hydrogen and oxygen are not usually found as atoms – they are more often found as molecules of H_2 and O_2, so we use these symbols in the equation:

$$H_2 + O_2 \rightarrow H_2O$$

During a chemical reaction, the reactants are broken up and the atoms are rearranged to form new substances. The atoms cannot be 'lost' or destroyed, and no new atoms can be formed. So, the number of atoms in the products of a reaction must always be equal to the number of atoms in the reactants. In other words, the equation must *balance*.

➡ Figure 8.10
The atoms in our equation can be represented as circles.

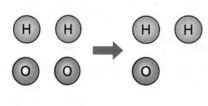

Is the equation we have written balanced? Look at Figure 8.10. The diagram shows that we are one oxygen atom short in the product. Scientifically, this cannot be the case. The number of atoms on the left-hand side of the equation must equal the number of atoms on the right-hand side.

➡ Figure 8.11
To form water molecules, we need two hydrogen molecules for each oxygen molecule.

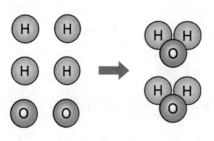

In reality, one oxygen molecule (reactant) combines with two molecules of hydrogen to form two molecules of water. Figure 8.11 shows this as a diagram.

The balanced equation for this reaction includes the numbers of molecules reacted and the numbers of molecules produced. So we write:

$$2H_2 + O_2 \rightarrow 2H_2O$$

The number in front of a formula shows you how many molecules of that substance are involved in the reaction. If there is no number then there is only one molecule.

When you write a chemical equation, you have to balance the numbers of atoms on the two sides of the equation.

Another example will help to make this clearer.

When zinc reacts with hydrochloric acid, it produces hydrogen gas and a salt called zinc chloride:

$$\text{zinc} + \text{hydrochloric acid} \rightarrow \text{hydrogen} + \text{zinc chloride}$$

This can be written as a chemical equation:

$$Zn + HCl \rightarrow H_2 + ZnCl_2$$

If you look at the atoms in the reactants and the atoms in the products, you can see that this equation is not balanced.

● There is only one atom of hydrogen (H) in the reactants and there are two in the products (H_2).
● There is only one atom of chlorine (Cl) in the reactants and there are two in the products (Cl_2).

We cannot change the chemical formulae of the molecules, so we have to balance the equation by changing the numbers of some of them.

In this example, we need more reactants because the extra atoms are both on the product side of the equation. It would not help to have more zinc, because those atoms are already balanced. It would help to have more HCl because it is the hydrogen and chlorine atoms that are unbalanced. If we use two molecules of HCl, the equation will balance:

$$Zn + 2HCl \rightarrow H_2 + ZnCl_2$$

Remember, the number 2 refers to both the H and the Cl atoms in hydrochloric acid.

Activity 8.4 Balancing equations

Write balanced equations for the following reactions.
1 magnesium + oxygen → magnesium oxide
2 magnesium + chlorine → magnesium chloride
3 sodium + chlorine → sodium chloride
4 zinc oxide + hydrochloric acid → zinc chloride + water
5 calcium + nitric acid → calcium nitrate + hydrogen

Chapter summary

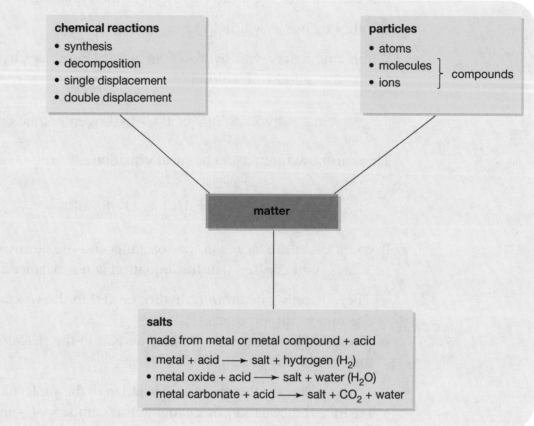

chemical reactions
- synthesis
- decomposition
- single displacement
- double displacement

particles
- atoms
- molecules
- ions } compounds

matter

salts

made from metal or metal compound + acid
- metal + acid ⟶ salt + hydrogen (H$_2$)
- metal oxide + acid ⟶ salt + water (H$_2$O)
- metal carbonate + acid ⟶ salt + CO$_2$ + water

Revision questions

1 Write a sentence to define each of the following terms.
   a) atom   b) molecule   c) ion   d) particle   e) covalent bond

2 Copy these sentences and complete them by filling in the gaps.
   a) When a metal reacts with a dilute _____ it produces a
      _____ and _____ gas.
   b) When a metal _____ reacts with a dilute _____ it produces
      a salt and _____.
   c) When a _____ carbonate reacts with _____ it produces a
      salt, water and _____ gas.
   d) When an acid reacts with a base we have a _____ reaction.

3 Write a balanced chemical equation to show the reactants and products when:
   a) zinc is reacted with hydrochloric acid
   b) copper carbonate is mixed with sulphuric acid
   c) magnesium oxide is mixed with hydrochloric acid.

Metals and the reactivity series

↑ **Figure 9.1** Why are so many of the items we use every day made from metals?

Metals have physical properties that make them very useful. Metals also have chemical properties and these cause them to react with other elements and compounds in different ways. In Chapter 8, you saw that metals react with acids to form salts. In this chapter, you will investigate how metals react with other substances and find out how metals are arranged in a reactivity series according to their levels of reactivity.

As you work through this chapter, you will:

- revise the physical properties of metals and link properties to uses
- investigate how metals react at different speeds with oxygen, water and acids
- understand how metals are arranged in the reactivity series according to how quickly or slowly they react with other substances
- explain displacement reactions in terms of the reactivity series.

Unit 1 Metals and reactivity

You already know that most of the elements in the Periodic Table are metals. You also know that metals have certain properties that make them very useful. Look at the pictures in Figure 9.2 and read the information to remind yourself of some of the useful properties of metals.

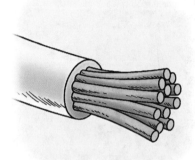

Copper is used to make electrical wires because it conducts electricity well.

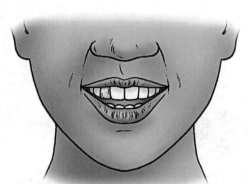

Gold and silver are used to make dental fillings because they are strong and they do not rust or corrode in the mouth.

Aluminium is used to make drinks cans because it doesn't rust and it is light.

Iron is used to make cast iron objects because it is strong and can be shaped easily.

Copper is used to make water pipes because it doesn't rust.

Gold and platinum are used to make jewellery because they are strong, shiny and durable.

↑ **Figure 9.2** How do people decide which metal to use for different purposes?

Think about how metals are used. Why don't we make water pipes out of sodium or even iron? Why do we coat iron with zinc to galvanise it? Why don't we use lead for modern water pipes? Why does magnesium burn brightly in a flame while iron and zinc burn steadily, and gold and silver don't burn at all?

When we choose a metal for a particular purpose, we must think about how it will react when it comes into contact with everyday substances like air, water and dilute acids. You already know that iron reacts with oxygen in air to form iron oxide (rust) and you have seen that magnesium reacts very quickly in dilute acids to form a salt and hydrogen gas.

In science, we use the word **reactivity** to talk about how quickly or slowly a metal reacts with other substances.

Some metals are very reactive – they react quickly and sometimes violently when exposed to air, water or acid. Sodium, for example, is so reactive that it has to be stored under a layer of oil to stop it coming into contact with the air. If it does, it reacts with the water and oxygen in the air and bursts into flames.

Other metals react so slowly that they are considered to be unreactive. For example, silver and gold don't corrode in air, don't dissolve or react with water, can be heated without bursting into flames and can be placed in dilute acids without forming salts.

Activity 9.1 **Comparing reactivity**

1 If a piece of silver jewellery is left exposed to the air for a very long time, it forms an extremely thin layer of oxide on its surface. If a piece of sodium is cut and exposed to the air, the shiny surface oxidises within seconds.
  a) What does this tell you about the reactivity of silver and the reactivity of sodium?
  b) Which of these metals is most reactive?

2 When iron oxidises, it forms rust. What can we do to prevent iron from rusting?

3 The metals in Group 1 in the Periodic Table, the alkali metals, are more reactive than the metals in Group 2, the alkaline earth metals. Think back to what you have learned about valence electrons and try to explain why Group 2 elements are less reactive than Group 1 elements.

Unit 2 Investigating reactions of metals

In Chapter 8, you did experiments to show that metals and metal compounds (oxides and carbonates) react with acids to form salts. But not all metals react at the same speed and some do not react at all.

Metals and dilute acids

If you put pieces of metals into dilute hydrochloric acid they react differently.

- Very reactive metals react violently – they produce bubbles of hydrogen gas and dissolve in the acid.
- Fairly reactive metals generally react more slowly and steadily, but some only react with concentrated acids.
- Less reactive and unreactive metals do not react at all, even in a concentrated acid.

Trudy and Gary investigated the reactions of different metals with a dilute solution of acid. It is extremely dangerous to put potassium, sodium or calcium into acids, so for these metals the pupils observed video clips of the reactions. Metals that are less reactive can be used safely at room temperature in the laboratory.

The table summarises Trudy and Gary's results. If the pupils found that a metal did not react in the dilute solution, their teacher placed it in concentrated acid to see if it reacted.

Metal	Reaction in dilute acid	Reaction in concentrated acid
potassium	violent reaction	
sodium	violent reaction	
calcium	violent reaction	
magnesium	reacts quickly but safely	
aluminium	reacts fairly quickly	
zinc	reacts fairly slowly	
iron	reacts slowly	
lead	no visible reaction	reacts slowly and safely
copper	very, very slow reaction	slow reaction
silver	no reaction	no reaction
gold	no reaction	no reaction

This jewellery is dirty with salts and oxides formed when the jeweller soldered pieces of silver together.

The dirty silver jewellery is left in a warm solution of dilute sulphuric acid.

The acid dissolves the dirt and the silver does not react with the acid.

↑ **Figure 9.3** Silver does not react with dilute sulphuric acid.

Metals and water

Very reactive metals react with cold water. Figures 9.4 and 9.5 show you what happens if you place sodium and potassium in water.

↑ **Figure 9.4** Sodium floats on water and fizzes as hydrogen gas is produced. The gas can make the sodium move across the water and it may burst into flames.

↑ **Figure 9.5** Potassium reacts violently with water, bursting into flames and producing hydrogen gas bubbles.

Calcium and magnesium sink in cold water. Calcium reacts slowly at first to produce hydrogen gas bubbles, but the reaction speeds up quite quickly. Magnesium reacts very slowly and produces a slow, steady stream of hydrogen gas bubbles. If magnesium is heated in steam, it reacts quickly and produces hydrogen gas.

Aluminium, zinc and iron sink in cold water and do not react at all. However, if these metals are exposed to steam they react slowly to produce hydrogen gas.

Metals and oxygen

Most metals react with oxygen. When they do, the product of the reaction is called a metal oxide.

Some metals react with oxygen at room temperature. When this happens, we see a layer of oxide on the surface of the metal.

Other metals only react with oxygen if they are heated. Reactive metals burn strongly, brightly and quickly when heated, leaving only powdered oxide.

Aluminium, zinc and iron glow and produce sparks if they are heated, but they don't normally burst into flames. Lead and copper do not burn when heated – they just acquire a layer of oxide.

Silver is not reactive, but it does form a layer of tarnish if it is left exposed to oxygen for a long time. Gold doesn't tarnish at all in air. Neither silver nor gold are changed by heating.

The reactivity series

You have seen that some metals are more reactive than others. Scientists have listed the metals in order from the most reactive to the least reactive. This list is called the **reactivity series**. The very reactive metals are at the top of the list. As you go down, the metals become less reactive. All metals can be included in the series, but our summary in Figure 9.6 shows only the metals you have worked with. Figure 9.6 also shows you how these metals react in acid, air and water.

Only the least reactive metals are found as elements in the ground. Unreactive metals like silver, gold and platinum are found in their pure state and they can be mined. The more reactive metals are only found combined with other elements in compounds. When these metals are extracted from the ground, they have to be processed to separate them from the other elements in the compounds. The less reactive metals like copper, iron and lead can be separated from their compounds easily. The more reactive metals, such as sodium and potassium, have to be separated from compounds using specialised processes.

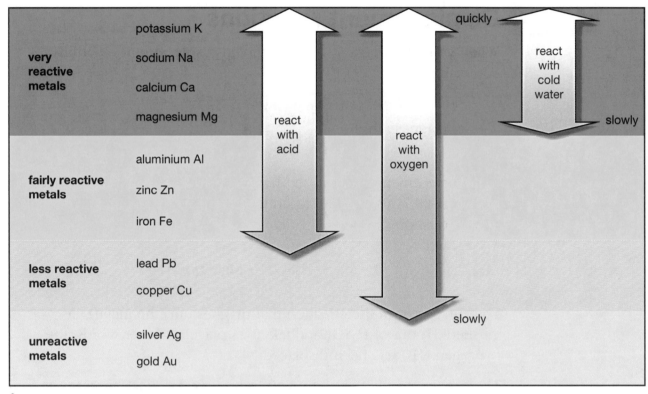

⬆ **Figure 9.6** We can arrange the metals in a reactivity series.

Activity 9.2   **Applying what you have learned**

1 A strip of unknown metal is held in a Bunsen burner flame.
It burns with a bright white flame and produces white smoke.
a) What metal could this be?
b) Do you think this metal reacts with oxygen? Why?
c) Explain why the unknown metal is not likely to be silver.

2 Lithium is a metal. It reacts violently with oxygen, cold water and dilute acid.
a) Where do you think lithium would fit in the reactivity series?
b) Would it be safe to place lithium in concentrated hydrochloric acid? Give a reason for your answer.

3 Why is gold mined as a pure metal, while zinc is not?

4 Why is copper a suitable material for water pipes?

5 The salt can be removed from sea water at special desalination plants, to produce fresh water for homes. The desalinated water is more acidic than ordinary household water. Do you think this will have an effect on copper water pipes? Give a reason for your answer.

Unit 3 Displacement reactions

When you prepared salts, you observed **displacement reactions**. For example:

$$\text{magnesium} + \text{sulphuric acid} \rightarrow \text{magnesium sulphate} + \text{hydrogen}$$

In a displacement reaction, the hydrogen from the acid is displaced by the metal. This is easier to understand if we use chemical formulae in the equation:

$$Mg + H_2SO_4 \rightarrow MgSO_4 + H_2$$

You can see from the equation that magnesium (the metal) has 'pushed' H_2 out of H_2SO_4 and taken its place. In other words, the hydrogen (H_2) has been displaced.

The reactivity series can help you to understand what happens in displacement reactions, and to predict which substance will be displaced in a reaction.

> In a displacement reaction, a more reactive metal will displace a less reactive metal from its compounds.

Look at the reactants and products in each of the displacement reactions shown in Figure 9.7. Can you see that, in each case, the metal that is higher in the reactivity series displaces the metal that is lower in the series?

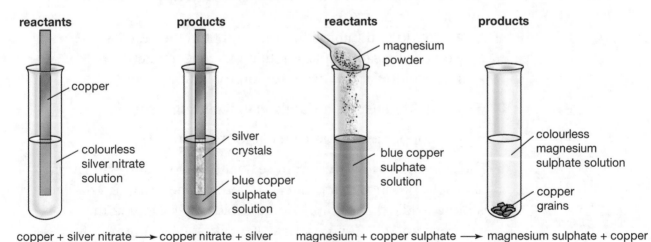

copper + silver nitrate ⟶ copper nitrate + silver   magnesium + copper sulphate ⟶ magnesium sulphate + copper

⬆ **Figure 9.7** In displacement reactions, the more reactive metal displaces the less reactive metal from its compound.

Experiment 9.1

Observing a displacement reaction

Aim

To observe what happens when an iron nail is placed in copper sulphate solution.

You will need:

- a test tube
- an iron nail
- a test-tube stand
- copper sulphate solution

Method

Set up your experiment as in Figure 9.8, and leave it for a week.

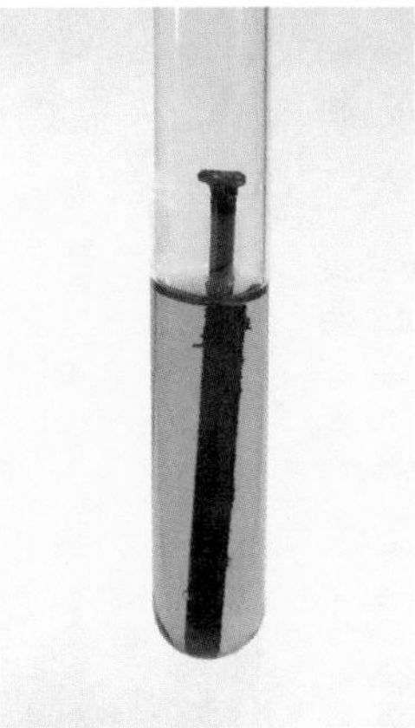

Questions

1 What colour was the nail to start with?
2 What colour was the solution in the test tube to start with?
3 What happened to the colour of the nail? Why?
4 What happened to the colour of the solution? Why?
5 Write an equation using chemical symbols to show what happened in this experiment.
6 How do you know that iron is more reactive than copper?
7 What would happen if you placed a copper wire in the solution instead of an iron nail? Why?

↑ Figure 9.8
How to set up your experiment

Activity 9.3   Tabulating reactivity

1 Copy the table. Complete it by adding ticks to show which metals would displace the metal from each salt solution. The first row has been done for you.

Salt	Metal					
	silver	copper	iron	lead	zinc	magnesium
copper sulphate			✓	✓	✓	✓
magnesium sulphate						
lead nitrate						
sodium sulphate						
iron chloride						
zinc nitrate						

2 Explain why none of these metals can displace the sodium from sodium sulphate.

Chapter summary

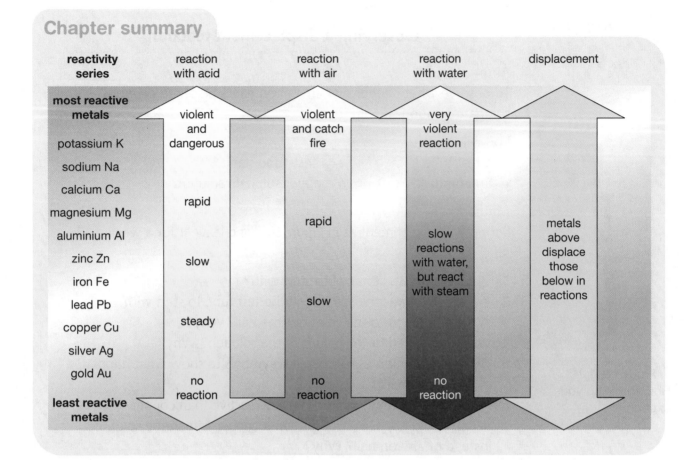

reactivity series	reaction with acid	reaction with air	reaction with water	displacement
most reactive metals	violent and dangerous	violent and catch fire	very violent reaction	
potassium K				
sodium Na				
calcium Ca	rapid			
magnesium Mg		rapid		
aluminium Al			slow reactions with water, but react with steam	metals above displace those below in reactions
zinc Zn	slow			
iron Fe				
lead Pb		slow		
copper Cu	steady			
silver Ag				
gold Au	no reaction	no reaction	no reaction	
least reactive metals				

Revision questions

1 Draw a labelled diagram to show what you could expect to see if you placed a strip of magnesium ribbon in a test tube of cold water.

2 Mercury is below copper in the reactivity series.
   a) Will mercury react with oxygen?   b) Will mercury react with dilute nitric acid?

3 Lithium reacts with cold water. It fizzes as it produces hydrogen gas but it does not catch fire. Would you place it above or below potassium in the reactivity series? Why?

4 What is a displacement reaction?

5 A pupil has three unknown metals. She labels them A, B and C.
   ● Metal B displaces A from a solution of its salts.
   ● Metal C displaces both A and B from a solution of their salts.
   a) Which metal is most reactive?   b) Which metal is least reactive?
   c) Write the metals in order of reactivity from most reactive to least reactive.

Chapter 10

More about chemical reactions

↑ Figure 10.1 When charcoal burns in air, it produces enough heat to cook meat.

In this chapter, you are going to learn more about chemical reactions. First you will learn about energy changes in reactions, looking at examples of reactions that produce heat and reactions that use heat. Then you will compare the rates at which different reactions take place and learn about the factors that can speed up or slow down chemical reactions.

As you work through this chapter, you will:

- describe energy changes in chemical reactions
- compare exothermic and endothermic reactions and give examples of each
- compare the speed at which different reactions take place
- investigate how concentration, heat, surface area and catalysts affect reaction rates.

Unit 1 Energy changes in chemical reactions

Whenever a chemical reaction takes place, there is a change in heat energy. There are two reasons for this:

- energy, in the form of heat, might be needed for the reaction to take place
- the reaction might produce, or give off, energy in the form of heat.

Endothermic reactions

When you have to add heat for a reaction to take place, the reaction is called an **endothermic reaction**. You can say that the reactants have taken in heat in order to form the products of the reaction.

↑ Figure 10.2 Cooking food is an example of an endothermic reaction.

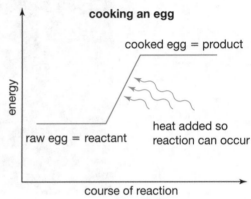

↑ Figure 10.3 Heat energy has to be added for the reaction to take place.

Other examples of endothermic reactions are:

- photosynthesis – plants take in energy from sunlight for the chemical reactions of photosynthesis to take place
- cold packs – when the chemicals in the cold pack mix, they take heat from their surroundings, so the pack feels cold.

Experiment 10.1

Observing an endothermic reaction

Aim

To observe temperature changes during an endothermic reaction.

You will need:
- 30 cm^3 citric acid solution ● 20 g baking soda ● polystyrene cup
- thermometer ● stirring rod

Method

1 Put the citric acid solution into the polystyrene cup, and measure its temperature. Remove the thermometer.

2 Stir the baking soda into the solution, and replace the thermometer right away. Record how the temperature changes.

Questions

Did the temperature increase or decrease? Can you explain why?

Exothermic reactions

When a reaction gives out (releases) heat, it is called an **exothermic reaction**.

↑ Figure 10.4 When a firework explodes, it gives out energy in the form of heat and light.

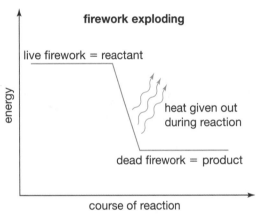

↑ Figure 10.5 Heat energy is released during the reaction and lost to the environment.

Other examples of exothermic reactions are:

● combustion (burning in air) – when any type of fuel burns in air, heat is released

● respiration – some of the energy released from food by respiration in body cells is given out as heat, which is why you get hot when you exercise (you are respiring more)

● hot packs – when the chemicals in the hot pack mix, they react together and produce heat, so the pack feels warm.

Activity 10.1 **Classifying reactions**

Say whether the following reactions are endothermic or exothermic.

1 a burning candle   2 eating a biscuit   3 sherbet fizzing in your mouth
4 a firefly glowing at night   5 an explosion   6 baking a cake
7 heating limestone to make lime

Unit 2 Rates of reaction

Some chemical reactions take longer than others. We say that chemical reactions proceed at different speeds – some are very fast, while others are very slow. The **rate** of a reaction is a measure of how fast it is going.

Figures 10.6 to 10.13 show you some chemical reactions that take place at different rates.

↑ **Figure 10.6** In this reaction, coal is processed in a factory to produce petrol.

↑ **Figure 10.7** In this reaction, the concrete takes some time to set.

↑ **Figure 10.8** In this reaction, the iron has rusted to form iron oxide.

↑ **Figure 10.9** Combustion is a chemical reaction that takes place quickly.

↑ Figure 10.10 The cut apple turns brown, as chemicals in the cells react with the air – if you keep the apple in a refrigerator, it doesn't go brown so quickly.

↑ Figure 10.11 Paint drying is a chemical reaction that can take place at different rates.

↑ Figure 10.12 Baking a cake is a chemical reaction.

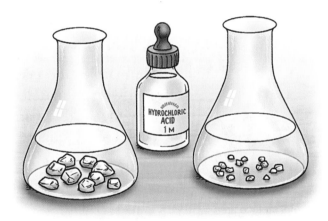

↑ Figure 10.13 Marble chips dissolve in acid. Which chips do you think will dissolve faster?

Activity 10.2 **Discussing reaction rates**

Look at the photographs and pictures on these pages.

1 Arrange the reactions in order from the one that takes longest to complete to the one that takes the shortest time.

2 Work in groups. Say why you think some reactions take longer than others.

3 Choose two slow reactions from your list. Can you think of any ways of speeding up the reaction rate? Can you think of any ways of making the reaction even slower?

Unit 3 Controlling reaction rates

It is often important to know the speed of a reaction and how to change it. In many industries, scientists need to control reactions to make sure they produce the correct amount of products in an economical, safe and environmentally acceptable way.

In a chemical reaction, particles in the reactants collide or bump against each other. If they collide with enough energy, they react and bond in different ways to form new substances – the reaction products.

The conditions in which a reaction takes place can affect the speed of the reaction. There are four main conditions that affect reaction rate.

1 Concentration

Increasing the number of molecules or ions in the reactants increases the chance of particles colliding. Generally, increasing the concentration makes a reaction go faster.

2 Particle size

A large marble chip takes longer to dissolve than several smaller ones. Powdered marble dissolves even faster than small chips. This is because the smaller particles have a greater surface area exposed to the acid. Look at Figure 10.14 to understand this.

Generally, the larger the surface area of solids in the reaction, the faster the reaction will go.

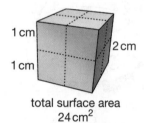

total surface area
24 cm^2

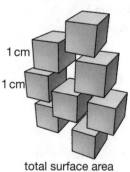

total surface area
48 cm^2

⬆ **Figure 10.14**
If you cut a cube with 2 cm sides into eight smaller cubes with 1 cm sides you increase the surface area that is exposed.

3 Temperature

Food in a refrigerator stays fresh for longer than food that is left outside the refrigerator. This is because the chemical reactions that cause food to decay take place faster at higher temperatures. This is true for most reactions – if you add heat to a reaction, it usually increases the reaction rate. If you cool the reaction, it takes place more slowly.

4 Catalysts

Catalysts are substances that speed up reactions when they are added in small amounts. The catalyst is unchanged in the reaction and it can be used again.

Catalysts made from platinum and rhodium are used in catalytic converters in car exhaust systems, to speed up the oxidation reactions that make poisonous exhaust gases less harmful. Catalysts are also used to make petrol in the catalytic cracking of crude oil, and to make margarine from vegetable oils by catalytic hydrogenation. Most industrial catalysts are made from metals such as platinum, palladium or rhodium. The enzymes that speed up digestion in the body are biological catalysts.

Experiment 10.2

Investigating reaction rates

Aim
To investigate the effects of concentration, particle size and heat on the rate of a reaction.

You will need:
- a conical flask
- a measuring cylinder
- a Bunsen burner
- a clock or stopwatch
- dilute hydrochloric acid
- marble chips of different sizes
- a thermometer

Method
Using the equipment listed, plan an investigation to see what effect particle size, temperature and concentration of acid have on the rate at which marble chips dissolve in hydrochloric acid. Your plan should include what you will do and how you will record your findings.

Show your plan to your teacher. Once it has been approved, carry out your investigation.

Questions
How do the following affect the reaction rate?
- changing the size of the marble chips
- heating the acid
- making the acid stronger or weaker

Activity 10.3   Answering questions about reaction rates

1 Why must chemical reactions be controlled?

2 What happens to a chemical reaction if:
  a) the concentration of the reactants is decreased?
  b) the reactants are heated?

3 Which will react faster with sulphuric acid – grains of copper sulphate or powdered copper sulphate? Why?

4 What is an enzyme?

5 Give an example of an industrial chemical reaction that is speeded up by catalysts.

Chapter summary

✓ Chemical reactions involve changes in energy.

✓ Endothermic reactions absorb or take in heat. Exothermic reactions give out heat.

✓ Reactions take place at different rates. The rate of reaction is a measure of the speed of the reaction – the higher the rate, the more quickly a reaction is completed.

✓ Reaction rates can be controlled by changing the conditions in which the reaction takes place.

✓ In general, reactions can be speeded up by increasing the concentration of reactants, the temperature at which the reaction takes place, or the surface area of solid reactants.

✓ Catalysts are substances that speed up reactions when they are added to the reactants. Enzymes are biological catalysts.

Revision questions

1 What is an endothermic reaction? Give an example.

2 What is an exothermic reaction? Give an example.

3 Explain why:
   a) an egg cooks faster at a higher temperature
   b) jelly sets faster if it is dissolved in less water
   c) a candle burns faster in a slight draught than in still air
   d) powdered chalk dissolves in vinegar faster than a lump of chalk does
   e) magnesium ribbon dissolves faster in a warmed solution of acid
   f) rennin (an enzyme) is added to milk to make cheese.

Density and pressure

⬆ Figure 11.1 Why do some things float on water whilst other things sink?

In this chapter, you are going to learn more about the properties and behaviour of solids, liquids and gases. First you will learn how to find the density of solids and liquids. Then you will learn about pressure and how it affects liquids and gases.

As you work through this chapter, you will:

- learn about density and how to measure it
- do experiments to determine the density of different solids
- understand what is meant by pressure and how it is measured
- find out about pressure in liquids and gases
- describe ways in which pressure can be put to practical use.

Unit 1 Density

Think about lead, rubber and wood. Which substance would you say is the heaviest?

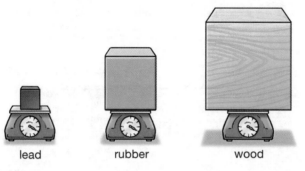

→ **Figure 11.2**
These blocks all have the same mass.

lead   rubber   wood

Look at the three blocks in Figure 11.2. They all have the same **mass**. However, they do not take up the same amount of space. Remember that the amount of space an object occupies is its **volume**. We can say that the wooden block has a much greater volume than the lead block.

To compare the heaviness of different materials, we need to compare pieces of the same volume. If we use blocks of the same volume, the mass of each material is different. Figure 11.3 shows you the mass of $1\,\text{cm}^3$ of different substances.

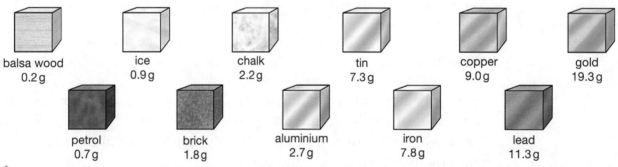

balsa wood 0.2 g   ice 0.9 g   chalk 2.2 g   tin 7.3 g   copper 9.0 g   gold 19.3 g

petrol 0.7 g   brick 1.8 g   aluminium 2.7 g   iron 7.8 g   lead 11.3 g

↑ **Figure 11.3**
These blocks all have the same volume but their masses differ.

The ratio between the mass and the volume of an object is called its **density**. You can think of density as the number of grams of a substance packed into each cubic centimetre (cm^3).

Determining density

You can work out the density of a substance using the formula:

$$\text{density} = \frac{\text{mass}}{\text{volume}}$$

Mass is measured in grams (g) and volume is measured in cubic centimetres (cm^3). This means the unit for density is grams per cubic centimetre (g/cm^3).

In Figure 11.3, gold has the highest density. Lead also has a high density. Water has a density of $1\,g/cm^3$, and most liquids (except mercury!) have densities near to this. Wood has a low density.

Every substance has a different density. Scientists can use the density of a substance to identify it. The tables show the densities of some common solids and liquids. You will not deal with the density of gases this year.

Solids	Density (g/cm³)
magnesium	1.7
aluminium	2.7
copper	9.0
gold	19.3
iron	7.8
lead	11.3
platinum	21.4
ice	0.9
cork	0.3
polystyrene	0.02

Liquids	Density (g/cm³)
water	1.0
petrol	0.7
orange juice	1.0
milk	1.0
cooking oil	0.9
methylated spirits	0.8
mercury	13.6

Floating and sinking

The density of objects affects whether they float or sink. For example, ice and oil are less dense than water, so they float in water. Aluminium and copper are denser than water, so they sink in water.

Measuring density of solids

You can find the density of any solid once you know its mass and its volume.

To find the mass of a solid, use a balance or mass meter.

The volume of a regular solid (cube or cuboid) can be found using the formula:

volume = length × breadth × height

The volume of an irregular solid can be found by lowering it into a liquid and measuring the displacement. You can do this using a measuring cylinder or a special piece of equipment called a Eureka can.

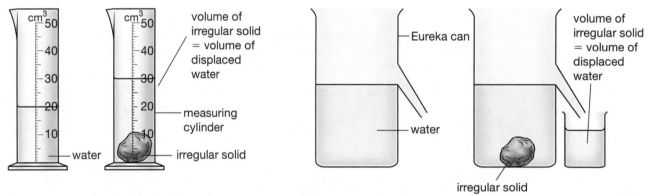

↑ Figure 11.4 Finding the volume of a solid

Measuring the density of liquids

Figure 11.5 shows you how to find the volume and mass of a liquid, using a measuring cylinder and a mass meter. Once you know these, you can use the formula to find the density of the liquid.

→ Figure 11.5
Remember to subtract the mass of the empty cylinder from the mass of the cylinder with the liquid, to find the mass of the liquid alone.

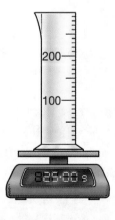

Experiment

11.1

Determining the densities of different objects

Aim

To measure and calculate the densities of different objects.

You will need:
- a small stone
- a small piece of wood
- a pencil eraser
- a measuring cylinder
- water and a mass meter
- a calculator
- a copy of the table below

Method

1 Predict which object is most dense and which is least dense. Record your predictions in your copy of the table.
2 Find the mass of the stone, wood and eraser. Record your measurements in the table.
3 Find the volume of the stone, wood and eraser. Record your measurements in the table.
4 Calculate the density of each substance.
5 Compare your measurements with your predictions. How well did you predict?

Object	Prediction	Mass	Volume	Density = mass ÷ volume
stone				
piece of wood				
pencil eraser				

Activity 11.1

Graphing densities

1 Arrange the solids in your results table from Experiment 11.1 in order from most dense to least dense.

2 Draw a bar graph to compare the densities of these solids.

3 a) Work out the density of the liquid in the cylinder in Figure 11.5.
   b) What liquid do you think it is?

Unit 2 Pressure

↑ Figure 11.6 It is easier to press the coin into the modelling clay when the coin is on its side.

Pressure tells you how concentrated a force is. When a large force is concentrated on a small area, the pressure is higher than when the same force is spread out over a larger area.

Figure 11.6 shows a coin being pressed into a lump of modelling clay. When the coin is placed flat on the clay, you can only press it in a little way. When the coin is on its side, you can press it deeply into the clay. This is because the force on the flat coin is spread out over a large area. The force on the edge is concentrated on a small area. So, although the same amount of force is supplied, the pressure on the edge of the coin is higher.

In science, pressure has a very exact meaning. It can be defined by the equation:

$$\text{pressure} = \frac{\text{force}}{\text{area}}$$

Force is measured in newtons (N) and area is measured in square metres (m^2). So pressure is measured in newtons per square metre (N/m^2).

High and low pressure

When an object rests on the ground, the weight of the object exerts pressure on the ground. The amount of pressure depends on the area of the object that is in contact with the ground. A force on a small area gives a higher pressure than the same force on a larger area. For example, a brick of weight 5 N exerts a different pressure depending on which face is in contact with the ground. You can see how this works in Figure 11.7.

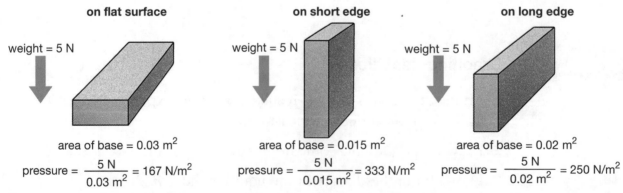

on flat surface

weight = 5 N

area of base = 0.03 m²

pressure = $\frac{5\,N}{0.03\,m^2}$ = 167 N/m²

on short edge

weight = 5 N

area of base = 0.015 m²

pressure = $\frac{5\,N}{0.015\,m^2}$ = 333 N/m²

on long edge

weight = 5 N

area of base = 0.02 m²

pressure = $\frac{5\,N}{0.02\,m^2}$ = 250 N/m²

↑ Figure 11.7 The weight of the brick does not change but the pressure changes depending on which face is in contact with the ground.

If you understand the idea of pressure, you can explain many things from everyday life. Look at the examples in Figure 11.8 to see how we make use of high and low pressure in daily life.

You can push a drawing pin into a surface because the point is a small area under high pressure.

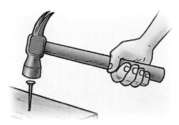

The force of the hammer is concentrated on the point of the nail.

A doctor only needs a small force to give a patient an injection.

The edge of a sharp knife has a small area, so it cuts easily when you apply pressure.

The broad feet of the camel spread its weight over a larger area and reduce the pressure on the sand.

Studs reduce the area of the boot in contact with the ground – the player exerts a greater pressure and so gets a better grip.

Narrow heels have a small area, so the person exerts more pressure on the floor.

Tractors are heavy. Their large tyres reduce the pressure they put on the soil.

When you walk on soft sand, your whole foot makes contact with the ground so you exert less pressure.

When you walk on stones your feet touch a smaller area and they get sore.

↑ Figure 11.8 Pressure depends on the force that is pressing down and the area that it is pressing on.

Activity 11.2 Calculating pressure

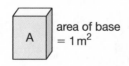
area of base = 1 m²

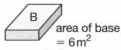
area of base = 6 m²

↑ Figure 11.9

1 In Figure 11.9, both blocks have a weight of 18 N.
   a) Which block exerts the most pressure on the ground? Why?
   b) Calculate the pressure exerted by each block.

2 A car of weight 6000 N has four wheels. The total area of contact with the road is 0.1 m². What pressure does the car exert on the road?

Unit 3 Pressure in liquids and gases

You have seen that a solid object exerts pressure through the area in contact with the ground. Liquids and gases behave differently. In liquids, the pressure acts on the bottom of the container but it also acts on the sides and on any objects that are put into the liquid. We can say that a liquid exerts pressure in all directions. Gases also exert pressure in all directions. These differences are largely due to the way that particles are arranged in solids, liquids and gases.

solid

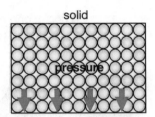

liquid

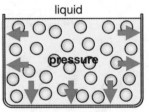

gas

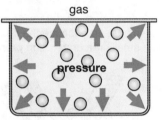

↑ Figure 11.10 The arrangement of particles means that liquids and gases exert pressure in all directions.

Pressure in liquids

Under water, you can feel the pressure on your body increase as you get deeper. This is because the weight of liquid above you increases as you go deeper. We say that the pressure in a liquid increases with depth. You can do Experiment 11.2 to demonstrate this.

Experiment 11.2

Investigating pressure in liquids

Aim
To show how water pressure increases with depth.

You will need:
● a large plastic bottle ● a sharp object ● water ● a ruler

Method
Work in pairs and do this experiment outside. Follow the steps in Figure 11.11 to set up your experiment.

Make three holes in the bottle at different heights.

Place the bottle on the ground next to a ruler.

Fill the bottle with water. Observe what happens.

↑ Figure 11.11

In liquids, pressure does not depend on the width of the container. The pressure at the bottom of the bottle in Experiment 11.2 is due to the weight of the water above it. A wide bottle would have a greater weight of water to support, but the force would be spread over a larger area. So in fact the pressure is the same at the bottom of a wide container as it is at the bottom of a narrow container, as long as they are filled to the same level with water.

In liquids, pressure also depends on the density of the liquid. Methylated spirit is less dense than water. If you did Experiment 11.2 with methylated spirit, the pressure on each jet of liquid would be less.

Pressure in gases

The Earth's atmosphere exerts pressure in a similar way to a liquid. Air pressure acts in all directions and the pressure at lower levels of the atmosphere is higher than at upper levels, because of the weight of air pressing down on it.

At sea level, air exerts a pressure of about $100\,000\,\text{N/m}^2$. This is the same pressure as exerted by the weight of four buses, and yet we don't feel the air pressing down on us.

We do not feel the effects of air pressure because we have air inside our bodies. The air inside our bodies exerts pressure outwards and this balances the inward pressure exerted by the atmosphere. If the air were removed from our bodies, we would be crushed by the air around us.

Activity 11.3 **Summarising what you have learned**

Copy and complete the mind map in Figure 11.12 to summarise what you know about pressure in solids, liquids and gases.

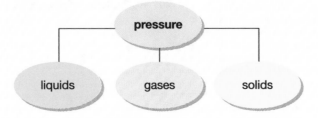

→ Figure 11.12

Unit 4 Making use of pressure

The way that liquids and gases behave under pressure allows us to use them in different ways. Look at the pictures and read the information carefully.

↑ Figure 11.13 Hydraulic lifting machines use liquid under pressure to exert a large force.

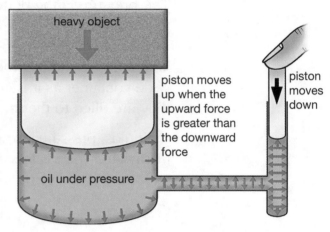

↑ Figure 11.14 How a hydraulic lifting machine works

heavy object

piston moves up when the upward force is greater than the downward force

piston moves down

oil under pressure

↑ Figure 11.15 Planes are able to fly as a result of air pressure.

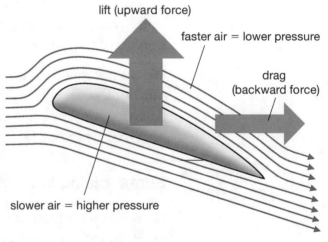

lift (upward force)

faster air = lower pressure

drag (backward force)

slower air = higher pressure

↑ Figure 11.16 Differences in air pressure create 'lift' and allow the plane to fly.

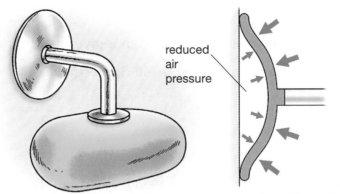

→ **Figure 11.17**
Air pressure allows us to use rubber suckers to stick things to surfaces.

the soap attachment is fixed to the wall with a sucker

reduced air pressure

cross section of the sucker

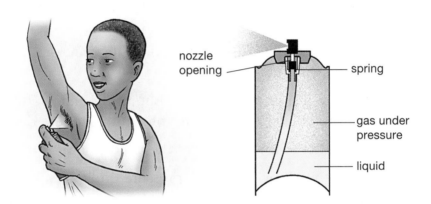

→ **Figure 11.18**
Aerosol sprays contain gas which is at a higher pressure than air. When you push the nozzle, the gas pushes the liquid out through the hole in the nozzle.

nozzle opening

spring

gas under pressure

liquid

Activity 11.4   Giving scientific explanations

Discuss these questions in groups. Use what you know about pressure to find the answers.

1 Why are dam walls thicker at the bottom than they are at the top?

2 Why do your ears sometimes pop when you drive up a high hill or travel in an aircraft?

3 Why is it difficult to push a ball under the surface of water?

4 Why can't you push the plunger into a syringe if your finger is over the opening at the front of the syringe?

5 The braking system in a car uses hydraulics. How does this work?

Chapter summary

- Density is the amount of matter contained in a given volume of a substance.
- Density (in g/m^3) can be determined using the formula:

$$\text{density} = \frac{\text{mass}}{\text{volume}}$$

- Pressure is the amount of force acting on a surface.
- Pressure (in N/m^2) can be determined using the formula:

$$\text{pressure} = \frac{\text{force}}{\text{area}}$$

- A force on a small area produces a higher pressure than the same force on a larger area.
- Solids exert a force downwards on the surface they are on. Liquids and gases exert a force in all directions.
- In liquids, pressure increases with depth but it is not affected by the width of the container. Denser liquids exert a greater pressure than less dense liquids.
- Air pressure is higher at lower levels of the atmosphere.
- Hydraulic systems work by exerting a pressure on liquids.
- Aircraft can fly, suckers can stick to surfaces and aerosol cans can spray because of differences in air pressure.

Revision questions

Use each of the following terms in a sentence to explain its scientific meaning.

1 weight

2 surface area

3 density

4 volume

5 pressure

6 liquid pressure

7 atmosphere

Chapter 12 · Forces and movement

↑ Figure 12.1 What forces are at work in this amusement park ride?

You have already learned how forces slow things down, speed things up and stop things moving. Forces can also cause objects to change direction or shape. In this chapter you are going to learn more about the effects of forces that we cannot see. You will learn about friction and air resistance, and how these forces affect movement.

As you work through this chapter you will:

- revise some ideas about forces and movement
- describe friction and how it can slow down or speed up movement
- understand how air resistance can cause objects to move at different speeds.

Unit 1 **Ideas about forces and movement**

You have already learned about forces. You should remember that a force is a push, pull or twist that acts on an object. A force can cause an object to move or stop it from moving. Forces can also cause an object to change its speed, direction of movement, or shape.

Forces are measured in units called newtons (N), using an instrument called a spring balance or newton meter.

Twisting the clothes makes them change shape.

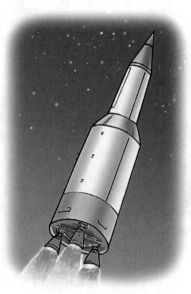

The rocket firing makes the spacecraft speed up.

A pulling force makes the door start moving.

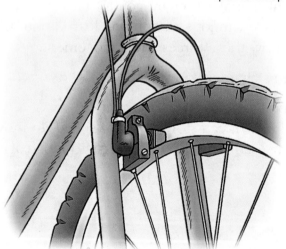

A pushing force on the brakes makes the wheel stop moving.

The force of the kick makes the ball change direction.

↑ Figure 12.2 Forces act on objects around us all the time.

Balanced and unbalanced forces

Forces act all the time, even if an object isn't actually moving. Think about a brick lying still on the ground. Gravity is acting on the brick. The Earth's gravity pulls everything in a downward direction, so the brick is being pulled towards the ground. At the same time, the ground exerts an upward push on the brick. These two forces balance each other out, so the brick does not move. We say the **net force** acting on the brick is zero newtons (0 N).

The brick will only move if the forces acting on it are unbalanced. If you pick up the brick, then you are exerting a pulling force that is stronger than gravity and the brick will move upwards in your hand. The net force acting on the brick is no longer 0 N, so it starts to move. Objects only move when unbalanced forces act on them.

When an unbalanced force acts on an object that is already moving, it makes the object speed up or slow down. When an object speeds up, we say it **accelerates**. If no unbalanced force acts on a moving object, its movement stays the same.

Activity 12.1 **Answering 'true or false' questions**

Say whether each statement is true or false. Rewrite each false statement to make it true.

1 Forces only act on moving objects.

2 We measure force in kilograms.

3 Forces can make objects change their speed and direction.

4 You need an unbalanced force to make an object change direction.

5 When the forces acting on an object are balanced, it will move at a constant speed in the same direction.

6 The scientific term for 'going faster' is speed.

7 In order for an object to move, the net force acting on it must be 0 N.

Unit 2 Friction

Friction is a force that acts against (resists) movement so it is called a resistive force. Friction happens when one surface comes into contact with another surface. For example, when you try to push a heavy object across the floor, there is friction between the object and the floor. Friction acts against the direction of movement. So, if you push the object forwards, friction acts against the pushing force. You have to push with a force stronger than the force of friction to make the object move.

push

friction

⬆ Figure 12.3 Friction is a resistive force that acts in the opposite direction to the direction of movement.

Smooth surfaces have less friction than rough surfaces. When two smooth surfaces come into contact, it is easy for them to slide over each other. When two rough surfaces come into contact, it is difficult for them to slide over each other.

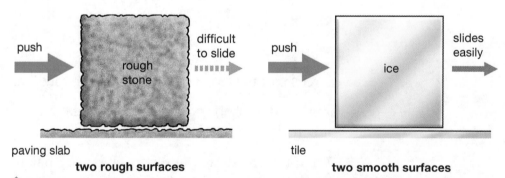

push

difficult
to slide

rough
stone

paving slab

two rough surfaces

push

slides
easily

ice

tile

two smooth surfaces

⬆ Figure 12.4 The size of the friction force depends on the roughness of the surfaces that come into contact.

The amount of each surface that comes into contact also affects how much friction there will be. Imagine if cars had no wheels – the whole bottom surface of the car would sit on the ground and it would be very difficult to move it. Wheels reduce the amount of surface that comes into contact with the ground, and allow the car to move more easily across the surface.

↑ Figure 12.5 The wheels help this skateboard overcome friction. The greased bearings reduce friction and allow the wheels to turn faster.

Reducing friction

When you want to make movement easier or faster, it helps if you can reduce friction between the surfaces. The most common way of reducing friction is to place a liquid between the two moving surfaces. This is called **lubrication**. For example, a layer of oil in car engines allows moving parts to slide over each other smoothly. The oil also helps reduce heat (caused by friction) and reduces wear and tear on the surfaces.

↑ Figure 12.6 The tread on the tyre helps grip the road. When the cyclist brakes, the friction between the brake pads and the wheels helps to slow them down.

Increasing friction

Friction can be useful. Without friction between the ground and our feet, we could not push forward and walk. Without friction between car wheels and the road, the wheels would spin and the car wouldn't move forward. Increasing friction allows us to start moving, prevent slipping or skidding, control the direction and speed in which we are moving, and stop moving.

Activity 12.2 **Applying your knowledge**

Use what you have learned about friction to write a few sentences explaining why:

1 drivers should slow down on wet roads

2 racing car tyres are smooth, but they are often wider than ordinary tyres

3 water runs down the slide at a water park all the time

4 ice skaters can move quickly and smoothly over ice.

Unit 3 Other resistive forces

When objects move through air and water, they too experience a type of friction. The effect is much less than friction between solids, but it can still affect movement. We call friction in air or water **resistance**.

Air resistance

What happens if you drop a sheet of paper and a crumpled-up ball of paper to the ground at the same time? Look at Figure 12.7.

When an object moves through the air, it pushes the air out of the way. As the air moves around the object, it pushes back on the object. The force of air against a moving object is called air resistance. Figure 12.8 shows how air resistance acts on a moving truck to slow it down.

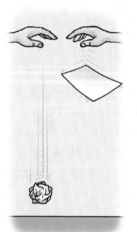

↑ Figure 12.7 The flat piece of paper falls more slowly because it has greater air resistance.

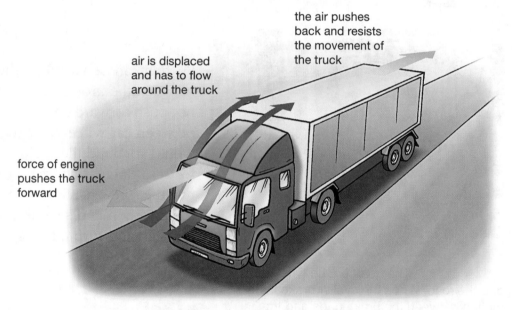

the air pushes back and resists the movement of the truck

air is displaced and has to flow around the truck

force of engine pushes the truck forward

↑ Figure 12.8 How a vehicle experiences air resistance

The size of the air resistance depends on the size and shape of the object. In Figure 12.7, the flat piece of paper moves more slowly because it has a much larger surface facing the air. The crumpled ball of paper moves more quickly because it has a much smaller surface pushing against the air.

Smooth, wedge-shaped objects move easily through air. These shapes are called **streamlined** shapes.

↑ Figure 12.9
Some ways of
changing the shape
of your car

Investigating air resistance

Aim

To find out how changing the shape of a vehicle affects its air resistance.

You will need:
- a toy car ● a ramp ● a stopwatch ● pieces of cardboard or paper
- sticky tape

Method

1 Time how long it takes for the car to run down the ramp.
2 Use the paper and cardboard to change the shape of the car.
   Figure 12.9 shows some examples of how you might do this.
3 Time how long it takes for the car to run down the ramp each time
   you change its shape. Record the shapes and the times taken.

Question

1 How can you make the car move faster?
2 How can you make the car move more slowly?
3 What does this tell you about the design of modern cars?

Overcoming and using air resistance

Air resistance tends to slow movement. Streamlined shapes are used to
reduce air resistance when things need to move faster. Figure 12.10
shows some ways in which air resistance is minimised by streamlining.

↑ Figure 12.10 Streamlined shapes help to reduce air resistance.

Air resistance can also be useful. Some animals use air resistance to help them move. For example, spiders have hairy legs. When they drop from a height they spread out their legs. This increases their air resistance and slows them down so they land safely.

↑ Figure 12.11 A flying lizard can't actually fly – large skin flaps increase its air resistance so that it can glide on air.

Sky divers use air resistance to land safely when they jump from an aircraft. When the skydiver dips forward, head-first, there is less air resistance and she moves very quickly. If she spreads out her arms and legs, she increases her resistance and slows down. Finally, she opens her parachute. The large area of the parachute experiences lots of air resistance so she slows down and lands safely.

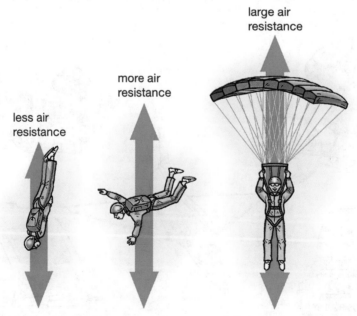

↑ Figure 12.12 Sky divers make use of air resistance to move quickly and land slowly.

Resistance in water

Water resistance is similar to air resistance. When an object moves through water, the water particles are pushed aside and they push back or resist the movement. However, water is denser than air, so the effects of resistance are greater. Water resistance is also called drag.

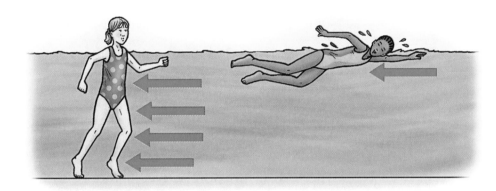

➜ **Figure 12.13**
Drag makes it more difficult to walk through water than to swim.

➜ **Figure 12.14**
Sharks move easily and quickly through water. Can you explain why?

Activity 12.3 **Choosing the correct words**

Choose the correct word from the brackets to make each sentence true.

1 Air resistance is the force of (gravity/friction) experienced by an object moving through air.

2 There is (more/less) resistance in water than in air.

3 Air resistance makes it (easier/harder) for cars to go fast.

4 An open parachute will (increase/decrease) air resistance.

5 Resistance in water is sometimes called (upthrust/drag).

6 Resistance can be (reduced/increased) by using a streamlined shape.

Chapter summary

✓ A force is a push, pull or twist that can make an object move or stop it from moving.

✓ Forces can also change the speed, direction and shape of objects.

✓ Only an unbalanced force will result in a change in movement.

✓ Force is measured in newtons using a spring balance or newton meter.

✓ Friction is a resistive force that opposes movement in solids.

✓ Friction is experienced when two surfaces move over each other.

✓ There is less friction between smooth surfaces than between rough surfaces.

✓ Lubrication reduces the effects of friction.

✓ Friction is experienced to a lesser degree in air and water. In these substances, friction is called resistance.

✓ Resistance can be minimised by streamlining shapes.

✓ Resistance in water is greater than in air because water is more dense than air. Water resistance is also called drag.

Revision questions

Write the correct scientific word or term for each definition.

1 The action that causes movement.

2 A pair of forces that result in an overall net force of 0 N.

3 The type of force that is needed to begin or stop movement.

4 The unit in which force is measured.

5 The contact force that resists movement in solid objects.

6 General term for forces that act against movement.

7 The action of air on an object moving through it.

8 Another name for water resistance.

9 A shape that is designed to move easily through air or water.

Chapter 13

Electrostatics and charge

↑ **Figure 13.1** What is making this girl's hair stand on end?

You have learned about electrical circuits and how electricity flows in a circuit. In this chapter, you are going to learn more about static electricity and electrostatic charges.

As you work through this chapter, you will:

- do experiments to learn about static electricity
- investigate how objects become charged
- describe how like and unlike charges behave
- find out how to induce an electric charge
- find out about the dangers and uses of static charges.

Unit 1 **What is static electricity?**

You have probably already observed **static electricity** in your everyday life. When you comb dry hair with a plastic comb, you might hear a crackling sound and your hair might stand on end. When you off take nylon clothes in a dark room, you might hear crackling or see small sparks. When you drag your feet across a carpet and then touch another person or a door, you might get a slight shock. These things are all caused by static electricity. You feel the effects best in cool, dry conditions.

Experiment 13.1

Producing static electricity

Aim
To observe what happens when certain materials are rubbed together.

You will need:
- a plastic item (a comb, ruler or plastic pen will work)
- pieces of woollen, nylon and cotton fabric
- some small pieces of paper

Method
1 Rub one end of the plastic item firmly and quickly against the woollen fabric.
2 Hold the end that you rubbed above the small pieces of paper, without touching them.
3 Repeat this using the nylon fabric, and then the cotton fabric. Record any differences in the results.

Questions
1 Does the plastic rod attract or repel the paper?
2 Which fabric produces the biggest effect on the plastic and the paper?
3 Try bringing the cloth close to the paper. Does it have any effect?

In Experiment 13.1, the plastic item attracts the pieces of paper. This happens because you produced static electricity by friction (rubbing the two surfaces together).

When you rub the plastic on the cloth, the item becomes electrically charged. One item gains negatively charged electrons from the other. When the plastic and cloth are separated, the excess negative charges stay in place. We say that the charges are static. The word 'static' means staying the same, or at rest. Static electricity is therefore electricity that stays in one place.

You can also produce electrostatic charge by combing dry, clean hair with a plastic comb, or by rubbing a balloon against your clothing.

Positive and negative charges

You already know that atoms contain protons and electrons. Protons have a positive charge and they do not move from atom to atom. Electrons carry a negative charge and they can be shared or transferred between atoms. (If you need reminding, look back at Chapter 7.)

- When an object loses electrons, it has more protons (+) than electrons (−) overall, so it becomes positively charged.
- When an object gains electrons, it has more electrons (−) than protons (+) overall, so it becomes negatively charged.

If you rub a polythene rod against a duster, the rod gains electrons and becomes negatively charged. If you rub a perspex rod against the duster, the rod loses electrons and becomes positively charged. Figure 13.2 shows you what happens if you bring the charged polythene and perspex rods together, and what happens when two polythene or two perspex rods come together.

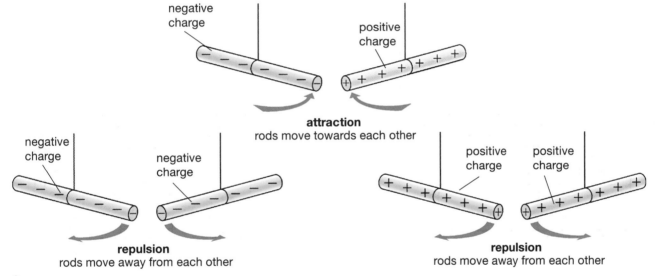

↑ **Figure 13.2** Opposite charges attract each other. Like charges repel each other.

Activity 13.1 Answering questions about electrostatics

When a plastic ruler is rubbed against a duster, electrons are transferred from the duster to the ruler.

1 What force causes the electrons to be transferred?

2 Which object becomes positively charged?

3 Which material becomes negatively charged?

4 Will the ruler and the cloth attract or repel each other once they are charged? Why?

Unit 2 Conductors, insulators and induction

Why does a plastic ruler become charged when you rub it against a duster, while a metal one does not?

This happens because the plastic ruler is an **insulator**. It does not allow the charge collected by rubbing to travel through it. We say the material does not conduct a charge. The charge therefore stays on the surface as static electricity. Materials such as glass, rubber, wool, silk, plastics and wood do not conduct electric charges, so they are all insulators.

The metal ruler will not become charged when you rub it against a duster because the metal allows the charge to travel through it. We say that metals are able to conduct a charge. This is because the electrons in metals are free to move from atom to atom. Metals are good **conductors** of electric charges. The human body is also a good conductor of electric charge because the liquid in our cells is able to conduct.

Because the electrons in a conductor are free to move, a charged object is always attracted by a conductor. Figure 13.3 shows how this works.

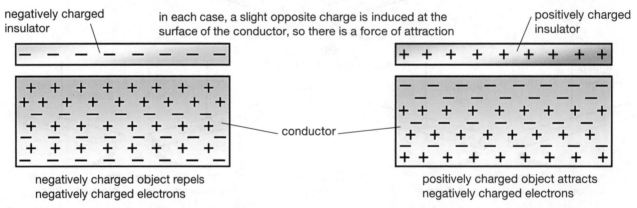

negatively charged insulator

in each case, a slight opposite charge is induced at the surface of the conductor, so there is a force of attraction

positively charged insulator

conductor

negatively charged object repels negatively charged electrons

positively charged object attracts negatively charged electrons

↑ Figure 13.3 A charged object is always attracted to a conductor.

Electric fields

You have already learned that objects with like charges repel each other and that objects with unlike charges attract each other. These forces of repulsion and attraction can act at a distance. You saw this in Experiment 13.1 when the pieces of paper were attracted to the charged plastic even though they were not touching it.

This tells us that there is an **electric field** surrounding a charged object. An electric field is the region of space around a charged object in which electric forces act. You will learn more about electric fields if you study science at higher levels.

Electrostatic induction

↑ **Figure 13.4**
The charged plastic induces a charge on the paper.

In Experiment 13.1, you brought a charged plastic item close to pieces of paper. The pieces of paper were not charged and yet they were attracted to the ruler. This suggests that the paper was carrying a charge opposite to the charge on the plastic ruler. However, this charge only lasts as long as the plastic ruler is close to the paper. When you move the plastic ruler away from the paper there is no longer any force of attraction. Figure 13.4 explains this.

In scientific terms, we say the plastic ruler has **induced** a charge on the surface of the paper. An induced charge is an electric charge produced in one object by the charge contained in another object. You can do a fun experiment to demonstrate electrostatic induction.

Experiment
13.2

Inducing a charge

Aim
To induce a charge on the classroom wall.

You will need:
● a balloon

Method
1 Blow up the balloon and tie it closed.
2 Put the balloon against the classroom wall. Does it stick to the wall?
3 Now rub the balloon against your school uniform to produce an electrostatic charge.
4 Put the charged balloon against the classroom wall. It should stick to the wall now.

↑ **Figure 13.5**
Why has this paper stuck to the balloon?

An uncharged balloon will not stick to the wall because there is no force of attraction between uncharged objects. However, once the balloon is electrostatically charged, it induces an opposite charge in the wall. This is strong enough to overcome the force of gravity on the balloon and it sticks to the wall.

Activity 13.2 **Explaining induction**

Look at Figure 13.5 carefully. Use what you have learned in this chapter to explain scientifically how you can make pieces of paper stick to the balloon.

Unit 3 Sparks, flashes and shocks

When electrostatic charge builds up on an object, it can cause sparks of electricity or flashes of light. A build-up of electrostatic charge can also be felt as an electric shock. Read the picture story in Figure 13.6, to see how this can happen.

This person is uncharged.

As he walks, the soles of his shoes lose electrons to the carpet and he becomes positively charged.

The positive charge is spread evenly around his body.

The positive charge on the man induces a negative charge on the door.

The negative charges jump from the door to his hand. He sees a spark and feels an electric shock.

⬆ **Figure 13.6** When charge builds up on a surface it can result in a spark or a shock.

⬆ **Figure 13.7** Lightning is caused by static electricity.

⬆ **Figure 13.8** This lighthouse has a metal conducting strip from the top to the ground.

Lightning

When air moves against large thunderclouds, electrons are transferred and a charge builds up on the cloud. The bottom of the cloud becomes negatively charged and the top becomes positively charged. At the same time, the surface of the Earth becomes positively charged. If the negative charge on the cloud builds up enough, a flash of electricity will spark from the cloud to the ground. We see this as forked lightning. Charges can also jump from the top to the bottom of a cloud in the form of sheet lightning.

A strip of conducting material has been attached to the lighthouse in Figure 13.8 and connected to the ground. We say that the lighthouse has been **earthed**. Earthing prevents lightning damage, because it allows excess electrons to be transferred to the Earth, so the object becomes uncharged or neutral. If lightning strikes the building, the electric charge travels down the conducting strip into the ground and the lighthouse remains uncharged.

Useful electrostatic charges

Photocopiers use electrostatic charges to make copies. The machine shines an image of the document you want to copy on to a positively charged drum. Where the light hits the drum, it conducts electricity and loses its charge. The dark parts, which are the image of the document, remain positively charged. The ink (toner) in the machine is negatively charged, so it is attracted to the positively charged parts of the drum. The drum then turns and presses the ink on to the paper, giving you a copy of the original document.

Electrostatic charge is also used to remove dust in industrial workplaces. The dust is attracted by an oppositely charged object and removed from the working area.

Activity 13.3   Applying your knowledge

1 Aircraft have special tyres that can conduct electric charge.
   a) Why would they need this?
   b) How do conducting tyres help prevent fires and explosions when a plane is on the ground?

2 Some cars have special conducting 'tails'. These are normally lengths of chain that hang down from the back bumper and touch the ground. Suggest why a car would need this.

3 Oil tankers are made from special materials so that people on board do not generate static electricity when they walk around. Explain why this is necessary.

Chapter summary

✓ Static electricity can be produced by the force of friction when you rub two materials together.

✓ Electrostatic charges can be positive or negative. The object that loses electrons becomes positively charged, while the object that gains electrons becomes negatively charged.

✓ Like charges repel each other but opposite charges attract each other.

✓ A charged object can induce an opposite charge in an uncharged material.

✓ Conductors are materials that conduct charge. Metals are good conductors.

✓ Insulators are materials that do not conduct charge. Glass, rubber, plastics and wood are all good insulators.

✓ When a charge builds up on an object it can cause sparks, flashes or electric shocks.

✓ Lightning is caused by static charges that build up in clouds.

✓ Static electricity is useful in photocopiers and for dust extraction in industry.

Revision questions

1 How do you produce static electricity?

2 Give three examples of materials that can be electrostatically charged.

3 What do we call materials that can be electrostatically charged?

4 Name three good conductors of electric charge.

5 Make a labelled sketch to show what happens when two positively charged rods are brought close together.

6 How can you tell whether two charged materials carry like charges or unlike charges?

7 Draw a simple, labelled diagram to show how a photocopier uses electrostatic charges.

8 Give two examples where electrostatic charge can be dangerous. Say what can be done to reduce the danger in each example.

Electricity

↑ **Figure 14.1** These wires are carrying electric current from a power station.

In Chapter 13, you learned about electricity at rest. In this chapter, you are going to learn more about what happens when electrical charges move in a flow called an electric current. You will work with electrical circuits and learn how to measure current. You will also investigate what happens to the current when you connect circuit components in different ways.

As you work through this chapter, you will:

- revise some of your earlier work on electric circuits
- define current and learn how it is measured
- design and build series and parallel circuits
- find out how current divides in parallel circuits
- investigate the effects of changing and adding components to circuits.

Unit 1 Electric current

You have already learned about electric circuits. Figure 14.2 will remind you of what you know.

When the switch is open or parts of the circuit are disconnected, the circuit is open and the bulb will not light up.

Symbols

$\dashv\vdash$ cell

$\dashv\vdash\vdash$ battery

switch open

switch closed

$\otimes$ bulb

―――― conductor

When the switch is closed and all the parts are connected, the circuit is closed and the bulbs light up.

➡ **Figure 14.2** You should remember how to draw a circuit using symbols.

When you built circuits like those in Figure 14.2, you saw that the bulb only lights up when the circuit is closed. In other words, the bulb only lights up when there is a complete path for electrical charges to flow through the circuit.

Current is a flow of charge. When no current flows in a circuit, the charges or electrons in the conductor can move freely in all directions. When the current flows, each charge is pushed forward and it bumps into the next one – this charge then bumps the next and pushes it forward. The movement results in a flow of electrons in the same direction round the circuit.

In reality, the negative terminal of a cell gives up electrons, which flow round the circuit towards the positive terminal. But the convention is to show the flow of current in a circuit using arrows from the positive to the negative terminal of the cell or battery.

Once the current starts flowing in a circuit, the same current keeps going round and round without getting used up.

You cannot see current, but you can tell that it is flowing when bulbs in the circuit light up. You can also measure current.

Measuring current

You can think of current as the rate at which electrons flow through a conducting wire. The SI unit of current is the ampere (A). This is usually shortened to amp.

Current can be measured using an instrument called an ammeter. Ammeters are specially designed so they do not affect the amount of current (they have a very low resistance).

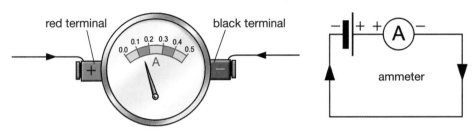

↑ **Figure 14.3** An ammeter is used to measure current.

An ammeter has a positive and a negative terminal. These are normally different colours and they are marked + and −. When you connect an ammeter in a circuit, you must connect it so that the positive terminal of the ammeter is closest to the positive terminal of the cell. If you connect it the other way round, you will damage the ammeter.

← **Figure 14.4** Most modern ammeters are combined with other instruments in a multimeter, which can also measure voltage and resistance.

Activity 14.1 **Using scientific terms**

Copy and complete these sentences by filling in the correct scientific term from the box. You can use a term more than once.

positive
negative
battery
circuit diagram
amperes
components
charge
ammeter

1 A _____ is two or more cells connected together.
2 The different pieces of equipment in a circuit are called _____.
3 Each component is shown by a symbol in a _____.
4 An electric current is a flow of _____ around an electrical circuit.
5 Current is measured in _____ by an _____ connected in the circuit.
6 The conventional flow of current in a circuit is _____ to _____.
7 When you connect an ammeter in a circuit, the positive terminal should be closest to the _____ terminal of the cell.

Unit 2 Series and parallel circuits

Components (like bulbs) can be connected in series or in parallel in a circuit.

In a **series circuit**, the components are connected in a row, or series. There is only one path for the current to flow.

In a **parallel circuit**, the components are connected next to each other, or parallel to each other. The current has to split to follow the different paths but it recombines once it has passed the parallel components.

Experiment 14.1

Designing circuits

Aim

To connect components in series and in parallel.

You will need:
- four cells
- four bulbs
- connectors
- a switch
- an ammeter

Method

Work in groups. Build the circuits shown in Figure 14.5.

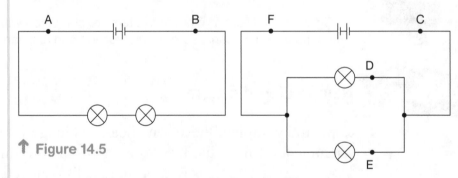

↑ Figure 14.5

Questions

1 Do the bulbs burn equally brightly in both circuits?
2 What happens in each circuit if you unscrew one bulb?
3 Use the ammeter to measure the current at points A, B, C, D, E and F in the circuits. What do you notice?
4 Design and draw a different series circuit and a different parallel circuit using only the components you have.

What happens in series and parallel circuits?

When components are connected in series, removing a component breaks the circuit and the current stops flowing. That is why unscrewing a bulb in the series circuit in Experiment 14.1 caused the second bulb to go out.

When components are connected in parallel, the current splits at the point where the circuit branches. When one component is removed, the current can continue to flow through the other branch, and the circuit is not broken. That is why unscrewing a bulb in the parallel circuit didn't cause the second bulb to go out.

Connecting the ammeter in the circuits shows that the current is the same all the way round a series circuit.

In a parallel circuit the current through all the branches adds up to the same as the current flowing from the cells. The current is shared by the components in parallel. The current leaving the parallel branches equals the current flowing into them.

Activity 14.2 Explaining and drawing circuits

1 Two identical bulbs are connected in series with three cells. Three ammeters are connected in the circuit to measure the current before and after each bulb. Each ammeter shows a reading of 1.5 A.
   a) Draw a circuit diagram to represent this circuit.
   b) Why do all three ammeters show the same reading?
   c) How will the ammeter readings change if the second bulb is removed from the circuit, and the wires are reconnected without it? Why?

2 The circuit diagram in Figure 14.6 shows two bulbs connected in parallel.

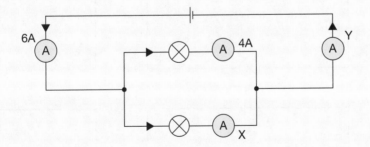

➡ Figure 14.6

   a) Write down the ammeter readings you would expect at points X and Y in this circuit.
   b) Redraw the circuit to show what the four ammeter readings would be if the same current flowed through each bulb.

Unit 3 Investigating circuits

In Experiment 14.1, why did some bulbs in your circuits shine more brightly than others? Bulbs (and other components) block or resist the flow of current in a circuit. The more a bulb resists, the less current will flow through it and the less brightly it will shine. So, for example, if there are two bulbs in series in a circuit, there will be twice as much **resistance** and the current will be halved.

As the number of bulbs and other components in a circuit increases, the resistance increases. This results in a smaller current.

Current can be increased by adding cells to the circuit. Two identical cells connected in series provide twice the current of one cell.

You are now going to build your own circuits, adding and removing components to see what effects your actions have on the flow of current.

Experiment 14.2

Investigating the effect of changing components

Aim

To investigate what happens when you change the number of components in a circuit.

You will need:
- two identical cells
- a switch
- two identical bulbs
- an ammeter
- connectors

Method

Use the components to build the series circuits shown in Figure 14.7.

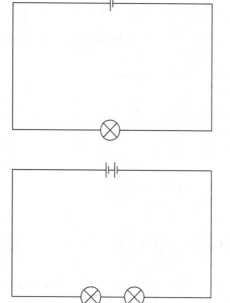
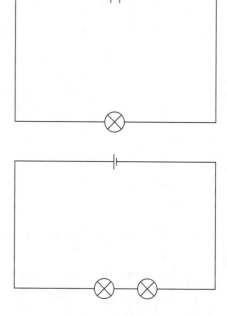

↑ Figure 14.7

Observe the differences in brightness of the bulbs in the different circuits.

Measure the current through components in the different circuits using the ammeter.

Copy and complete the table to summarise your results.

Table A – series circuits

Change	Effect on the bulbs	Effect on the current
adding more cells		
adding more bulbs		

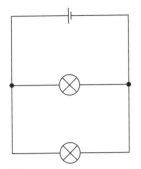

↑ Figure 14.8

Now build the parallel circuit shown in Figure 14.8.

Investigate what happens in a parallel circuit when you add or remove components.

Draw each circuit you build and note the changes you observe.

Copy and complete the table to summarise what you find out.

Table B – parallel circuits

Change	Effect on the bulbs	Effect on the current
adding more cells		
adding more bulbs		

Activity 14.3   Circuits in everyday life

1 Are the lights in a home connected in series or in parallel? How do you know?

2 Shaheed has two lights in his bedroom. The ceiling light is very bright. He uses it to light up the room when he is getting dressed in the morning and doing his homework at night. The bedside lamp is very dim. He uses it when he gets into bed at night to read.
   a) Which light blub do you think has the bigger current flowing through it when it is on? Why?
   b) Which light bulb has the greater resistance? How do you know this?

3 A string of fairy lights in a row are used for decoration. The lights are all on one wire, which is attached to a battery pack.
   a) Is this a series or a parallel circuit? Why?
   b) What is likely to happen if one bulb blows? Why?

Chapter summary

✅ Circuit diagrams are drawn using symbols to represent different components.

✅ An electric current is a flow of negatively charged electrons around a closed electric circuit.

✅ The conventional flow of current is shown from the positive terminal of the cell or battery to the negative terminal.

✅ Current is measured in amperes (A) using an ammeter.

✅ An ammeter has low resistance so it doesn't affect the flow of current when it is properly attached in a circuit.

✅ In a series circuit, components are connected in a row and there is only one path for the current.

✅ In a parallel circuit, components are connected parallel to each other and the current splits to flow through them.

✅ The current is the same at different points in a series circuit.

✅ The current entering a parallel circuit is equal to the sum of the current through the parallel branches.

✅ Adding cells to a circuit increases the current through the circuit.

✅ Adding components to a circuit increases the resistance in the circuit and makes it more difficult for current to flow.

Revision questions

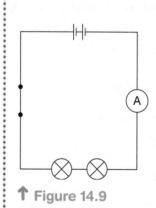

↑ Figure 14.9

Study the circuit diagram in Figure 14.9 and answer the questions.

1 What symbols have been used to show a bulb, a switch and an ammeter?

2 Is this an open or a closed circuit? How do you know?

3 Why does the bulb light up when the switch is closed?

4 Is this a parallel or a series circuit?

5 What is measured with an ammeter and what units are used?

6 What will happen to the brightness of the bulb if you add another cell in series in this circuit?

7 What happens in the circuit if you remove a bulb from it? Why?

Glossary words

A

accelerate   increase speed, when a force makes an object move faster

adaptation   a characteristic that makes something suitable to a particular environment

artificial selection   the process of breeding plants or animals with desired characteristics

atom   smallest part of an element

atomic number   the number of protons in the nucleus of an atom

B

biodiversity   the wide variety of plants and animals in an environment

bond   word used to describe how atoms are held together in a chemical compound

C

carnivore   an animal that eats the flesh (meat) of other animals

carpel   the female reproductive organ of a flower

catalyst   a substance that helps to speed up a chemical reaction – the catalyst is not used up during the reaction

cellulose   substance that makes up the stringy indigestible fibres found in the cell walls of plants

characteristics   features that are typical of a particular thing and that make it different from other things

chemical change   a change which results in new substances being formed

chemical formula   a way of representing the type and number of atoms in a compound using the symbols for the elements

chemical reaction   a change in which atoms are rearranged to form new chemical substances

chlorophyll   a chemical found in plants which gives them their green colour

chloroplast   a small structure in a plant cell that contains chlorophyll

conductor   a material that allows heat or electricity to pass through it

consumer   an animal that eats other animals or plants in an ecosystem

core electrons   an atom's inner electrons, which do not take part in chemical bonding, they are found in full shells

covalent bonding   chemical bonding in which atoms are held together by sharing one or more pairs of electrons

cross-pollination   pollination of one flower by pollen from another flower

current   a flow of electricity, the movement of electrons through a conductor

D

decomposers   bacteria and fungi which break down other substances

density   a measure of the mass of a substance by unit of volume

dispersal   the spreading of fruits or seeds of plants by agents such as wind, water or animals

displacement reaction   chemical reaction in which one substance pushes out another from a compound and takes its place

E

earthed   grounded, when the electricity supply is attached to a conductor that carries the current to the ground in the case of a fault

ecological pyramid   a diagram that shows the relationships between organisms in an environment – can show number, biomass or energy

ecosystem   all the plants, animals and non-living things in an environment

electric field   the area in which forces are exerted on an electric charge

electron configuration   describes the arrangement of electrons in an atom

electron   tiny particle outside the nucleus of an atom, negatively charged

embryo   a plant or animal in its early stages of development

endothermic reaction   a chemical reaction in which heat is absorbed as the reaction progresses

environmental variation differences in members of the same species caused by conditions in their environment

eutrophication excessive growth of algae in a water source as a result of pollution by organic matter

exothermic reaction a chemical reaction in which heat is given out as the reaction progresses

extinct having no living members of a species, died out

F

fertilisation the joining of male and female gametes during sexual reproduction

food chain a relationship between living things that depend on each other for food energy

food web a diagram showing the linked food chains in an ecosystem

friction the force that resists or works against the movement of an object

fruit the structure formed by the ovary of a plant once it has been fertilised

G

gamete a reproductive cell

germination the first stages in the growth of a seed as it grows into a small plant (seedling)

group column on the Periodic Table that contains elements with similar chemical properties

guard cells pairs of special cells that protect the openings (stomata) on the leaves of plants

H

herbivore an animal that eats only plant materials

I

indigenous found naturally in a particular area

induce cause to happen

inherited characteristics characteristics that are passed on from parents to their offspring

insulator a material which reduces or stops the movement of heat or electricity through it

ionic bonding chemical bonding in which oppositely charged ions attract each other

ion an atom or groups of atoms with an electric charge

isotopes atoms of the same element which have the same number of protons but different numbers of neutrons

L

lipid biological molecule, found in oils and fats

lubrication adding a substance (such as grease or oil) in order to reduce friction between surfaces

M

mass a measure of how much matter there is in an object

mass number total number of protons and neutrons in an atom

metallic bonding bonding in which positively charged metal atoms are held together by a 'sea' of negatively charged electrons

metalloid an element which is not obviously a metal or a non-metal but which has properties of both

micro-organism any living thing that can be seen only with a microscope

molecule a group of atoms joined together

N

net force the overall force acting on an object, when forces are balanced, the net force is zero

neutralisation reaction chemical reaction in which an acid and an alkali react to form water and a salt

neutron tiny particle in the nucleus of an atom, carrying no charge

nucleus tiny central part of an atom, containing almost all of its mass

O

omnivore an animal that eats both plants and other animals

P

parallel circuit an electrical circuit in which the current splits into two or more paths and then joins up again

period a row on the Periodic Table

photosynthesis the process by which plants turn sunlight into food

physical change a change that is easily reversed and in which the substances themselves do not change

pollen the powdery grains that contain the male gametes of plants

pollinate   transfer pollen from a male reproductive organ to a female reproductive organ (in plants)

pollution   an unwanted change in the natural environment, usually caused by human activity

population   the number of plants or animals of a particular species living in a habitat

pressure   a force spread over an area

producer   the plants that are at the start of every food chain

product   a substance formed in a chemical reaction

protein   biological molecule, component of food

proton   tiny particle in the nucleus of an atom, positively charged

pyramid of biomass   a diagram showing the amounts of living material (biomass) found at different levels in a food chain

pyramid of energy   a diagram showing the amounts of energy available at different levels in a food chain

pyramid of numbers   a diagram showing the numbers of organisms found at different levels in a food chain

R

rate   a measure of how quickly something happens

reactant   a starting substance for a chemical reaction

reactivity   the rate at which a substances reacts with other substances

reactivity series   a sequence of metals to show the rate at which they react with oxygen and with water

reproduction   making new versions of themselves, producing offspring

resistance   (electrical) a measure of how strongly a component opposes the passage of an electrical current

resistance   (force) a force that works against other forces, for example when air offers resistance against movement

respiration   using oxygen to get energy from food

S

salt   a chemical substance formed when a metal (or a metal compound) reacts with an acid

selective breeding   breeding plants and animals to produce offspring that have those characteristics of their parents that the breeder wants

series circuit   an electrical circuit in which the components are connected end-to-end

shell   region in which electrons orbit the nucleus of an atom

soluble   able to dissolve in a liquid

stamen   the male reproductive organ of a flower, consists of an anther on a thin stalk (filament)

starch   an insoluble biological molecule used to store food in plants

static electricity   energy that is built up when electrons are transferred from one object to another object, not flowing in a current

stomata   small pores found on the surface of leaves, mostly on the underside

streamlined   shaped to move easily through air or water

T

test   scientific investigation

transport   move from one place to another

trophic levels   the positions that different organisms occupy in food chains, some organisms may feed at more than one trophic level

V

valence electrons   the electrons found in the outermost shell of an atom, which take part in chemical bonding

variation   small differences between members of the same species

volume   the amount of space inside an object or occupied by a solid object, measured in cubic units